T0381182

Responsible AI in Practice

A Practical Guide to Safe and Human AI

Toju Duke
Paolo Giudici

Apress®

Responsible AI in Practice: A Practical Guide to Safe and Human AI

Toju Duke
Bedrock AI, London, UK

Paolo Giudici
University of Pavia, Pavia,
Italy, Via San Felice 7, 27100

ISBN-13 (pbk): 979-8-8688-1165-4
ISBN-13 (electronic): 979-8-8688-1166-1
https://doi.org/10.1007/979-8-8688-1166-1

Managing Director, Apress Media LLC: Welmoed Spahr
Acquisitions Editor: Shaul Elson
Development Editor: Laura Berendson
Coordinating Editor: Gryffin Winkler

Cover image by Katrin Bolovtsova from Pexels (https://www.pexels.com/)

Distributed to the book trade worldwide by Springer Science+Business Media New York, 233 Spring Street, 6th Floor, New York, NY 10013. Phone 1-800-SPRINGER, fax (201) 348-4505, e-mail orders-ny@springer-sbm.com, or visit www.springeronline.com. Apress Media, LLC is a California LLC and the sole member (owner) is Springer Science + Business Media Finance Inc (SSBM Finance Inc). SSBM Finance Inc is a **Delaware** corporation.

For information on translations, please e-mail booktranslations@springernature.com; for reprint, paperback, or audio rights, please e-mail bookpermissions@springernature.com.

Apress titles may be purchased in bulk for academic, corporate, or promotional use. eBook versions and licenses are also available for most titles. For more information, reference our Print and eBook Bulk Sales web page at http://www.apress.com/bulk-sales.

Any source code or other supplementary material referenced by the author in this book can be found here: https://www.apress.com/gp/services/source-code.

If disposing of this product, please recycle the paper

Table of Contents

About the Authors

Toju Duke, Recognized as one of the top women in AI, Toju is an entrepreneur, speaker, author, thought leader, and advisor on Responsible AI, with over 19 years of experience spanning advertising, retail, nonprofit, and tech. She spent a decade at Google, where her last three years as a Responsible AI Program Manager focused on leading Responsible AI initiatives across Google's product and research teams, with a focus on large-scale models and Responsible AI processes. Toju is the founder and CEO of Bedrock AI, an AI product and services company dedicated to Responsible AI principles. She is also the founder of Diverse AI, a community interest organization with a mission to support and champion underrepresented groups to build a diverse and inclusive AI future. In 2023, she published *Building Responsible AI Algorithms: A Framework for Transparency, Fairness, Safety, Privacy, and Robustness* (Apress).

Paolo Giudici is Professor of Statistics at the University of Pavia and Professor of Machine Learning at the European University Institute. He is author of more than 200 scientific publications, with an h-index of 52 (Google Scholar), 39 (Scopus), and 35 (Web of Science). The publications propose statistical learning models that can measure the opportunities and risks of artificial intelligence and financial technologies and improve their quality and safety. He is Chief Editor of the scientific journal *Statistics* (Taylor and Francis) and Editor of *Artificial Intelligence in Finance* (Frontiers) and of *International Journal of Data Science and Analytics* (Springer). Paolo has coordinated 14 funded scientific projects, among which are the European Horizon 2020 projects "PERISCOPE: Pan-European Response to the Impacts of COVID-19 and future Pandemics and Epidemics" (2020–2023) and "FINTECH: Financial Supervision and Technological Compliance" (2019–2020). The projects have supported the research activity of 19 PhD students and of 17 postdoc researchers. He has been a research fellow at the Bank for International Settlements and a research expert for the European Commission, the European Insurance and Occupational Pensions Authority, the Italian Ministry of Development, the Bank of Italy, and the Italian Banking Association. He has been a board member of the Credito Valtellinese bank (2010–2018), and he is an honorary member of the Italian Financial Risk Management Association. He is also an elected fellow of the International Statistical Institute (ISI), and a member of the Institute of Mathematical Statistics (IMS), the Association for Computing Machinery (ACM), the European Network for Business and Industrial Statistics (ENBIS), and the Italian Statistical Society (SIS).

Acknowledgments

This book is a result of extensive research and stakeholder management between researchers and industry professionals particularly from the statistical laboratory of the University of Pavia. Special thanks to the experts and regulators at the European Supervisory Digital Finance Academy of the European University Institute who helped to shape the rationale and ideas behind this book. We also cannot thank Golnoosh Babaei enough, who worked tirelessly on contributing to the case studies and Python code presented in this book.

Introduction

Autonomous cars, Internet of Things (IoT), smart homes, smart phones, smart cities, recommender systems, personalized recommendations, personalized health monitors, surveillance cameras, and so on and so forth. What do these all have in common? In today's world, the main underlying technology driving these systems is artificial intelligence (AI). When the term was introduced in 1956 by Alan Turing and a few other professors, there was no clear vision or prediction on how this technology would turn out. But one thing was sure – someday we'd have a technology that could not only mimic human intelligence but possibly surpass it. Ever since AI was introduced to our world, it's made incredible leaps and bounds that have wowed many, and we're constantly being introduced to a newer, more capable AI model on a daily basis.

These recent AI breakthroughs have changed the way we live, the way we think and interact with one another, our devices, and the world at large. We no longer conduct only "search" on the Internet, but we now have conversations by chatting with it. We don't only write emails, but we're able to have corrections, suggestions, and changes while doing it, with the additional option to simply ask our AI assistant to complete the task for us. Remember the days of Clippy, Microsoft's friendly-looking paper clip that couldn't do much, but was a great addition to Microsoft Office suite? Well, we're now equipped with numerous capable and efficient assistants that are so much more productive, efficient, and interactive than Clippy ever was.

With the recent advancements in AI technologies, particularly Generative AI which has not only taken the world by storm but has introduced the way we interact and use technology in so many different and new ways, there has been a steep increase in AI use and adoption over recent years. From multimodal AI systems such as text to image, to video, to audio, to voice, and robotics, we are seeing rapid development in one AI system that can achieve tasks from different modalities.

However, with this increased advancement of AI technologies comes an increase in concerns around its use, safety, transparency, trustworthiness, security, privacy, and reliability. There's also huge concerns on the amount of energy these systems consume. These valid concerns led to the quick introduction of regulation by several regulatory bodies and countries across the world, including Western countries and countries from the Global South. While there's been a good number of existing governing guidelines on AI development and use, there's little guidance on how to implement these ethical and responsible requirements during AI development, deployment, monitoring, and assessment.

We're helping to bridge this knowledge gap with the introduction of Responsible AI frameworks. Building on the framework introduced in *Building Responsible AI Algorithms* by Duke (2023), this book can be considered a sequel to this work. We've built an additional framework that goes a step further by providing statistical and programming code that can measure the various Responsible AI principles introduced in the new and additional framework. We've also ensured existing regulatory risk management frameworks are understood and referred to, enabling proper governance and risk management analysis in AI development workflows.

This work is born out of a genuine passion and desire to see more AI systems built with the safety and well-being of the end user in mind, so society and consumers can leverage the benefits of AI, while keeping the challenges and dangers associated with this technology at a bare minimum. We hope you gain the required insights and guidance from the extensive work outlined in this book – to build, test, and deploy AI systems safely which will benefit businesses and individuals alike and reduce the several harms that have already impacted so many human lives.

PART I

Introduction

CHAPTER 1

Responsible AI and AI Governance

Responsible AI

Responsible AI is a new and nascent field, and the term "Responsible AI" has been used interchangeably with the term "ethical AI" in recent years. In this chapter, we'll look at a brief history of Responsible AI and the factors influencing its emergence as a new and incredibly valuable part of AI development. We will also review new global AI regulation and policies from UNESCO to the EU AI Act and a few other global AI policies, highlighting the different principles in each policy and the importance of adopting the recommendations to build AI governance within your organization/project.

Responsible AI falls within the field of "ethical AI" and refers to the governance, responsible, social, societal, and legal aspects of AI. Due to the sporadic and unpredictable nature of AI applications which have resulted in several risks and harms perpetuated on various members of society, the need for more responsible development and deployment of AI has increased over the years, which also includes its use.

T. Duke and P. Giudici, *Responsible AI in Practice*,
https://doi.org/10.1007/979-8-8688-1166-1_1

In simple terms, Responsible AI can be defined as a body of work that ensures the development, deployment, and use of AI models and applications which have minimal risks and harms on every member of society, regardless of race, gender, belief, sexual orientation, age, and ability (disability).

The field of "responsible principles in AI" isn't new and has been in existence since the 1950s round about the time the term artificial intelligence was coined in 1956 by John McCarthy of the Massachusetts Institute of Technology (MIT) during the first academic conference on computer programming construction and the ability for machines to think – a concept introduced by Alan Turing, who is considered the father of AI and one of the first pioneers in the field.

Due to the issues stemming from AI applications in the past few years, which negatively affected users in various ways, organizations and companies turned to the philosophical academic research that originated in the 1950s, focused mainly on philosophical arguments ranging across several principles from fairness, transparency, accountability, privacy, safety, explainability, inclusivity, and sustainability.[1] These arguments encompass the field of AI ethics and include Responsible AI principles.

We've seen a steady growth of AI adoption over the years, and we're currently in an "AI summer," mainly driven by the recent introduction of generative AI – deep learning models that are able to generate high-quality content ranging from text, audio, video, and speech. The arrival of generative AI led to the AI arms race in 2022 where we saw a plethora of rapid AI development built on the transformer architecture (introduced by Google in 2017, a new neural network architecture for language models[2]) with the introduction of text to image generators, text to video, text to speech, text to image and video, and so many other multimodalities, kicked off by the famous ChatGPT.

We've also seen a steady and quite concerning increase in the number of human rights violations, psychological harms, data leakages, energy consumption, disinformation, social inequities including representational harms and biases, gender- and child-related violence, which also includes an increase in child sexual abuse material (CSAM), and threats to democracy and elections with a heightened increase in deepfakes. These are very concerning issues that cannot go ignored.

In *Building Responsible AI Algorithms* by Toju Duke,[3] a good number of real-life examples of AI's harms on various members of society were discussed. Duke also introduced a Responsible AI framework which offers solutions to mitigate identified risks and harms that arise as a result of AI technologies' outputs/results. We recommend reading *Building Responsible AI Algorithms* first before delving into the principles of this book, as it sets the foundation and cornerstone for Responsible AI principles and guidelines and provides some background on the risk management framework and measurement techniques outlined in this book.

To relay a few very recent examples where AI has produced harmful output to users, let's take a look at a couple of examples below.

The LAION-5B dataset, which stands for Large-Scale Artificial Intelligence Open Network, was introduced in March 2022, just before generative AI took the world by storm, by a German nonprofit organization – LAION. LAION develops datasets, tools, and models for machine learning (ML) research. Sourced from the Common Crawl web index, LAION-5B is a popular open source training dataset containing over 5.8 billion images used for image generation. It was used to train Stable Diffusion, an image generator introduced by Stability AI (a UK-based start-up), and other AI image models. LAION has been riddled with all sorts of risks ranging from copyright, privacy, and safety which has been covered by several media outlets.[4]

In April 2023, a German stock photographer discovered that his photos had been used to train the LAION-5B dataset and requested for the photos to be removed as they'd been used without his consent, knowledge, and any form of compensation, infringing on copyright laws (copyright is still a gray and debatable area with respect to AI and dataset training). On his request to the company to delete his pictures from the dataset, he was met with a $979 fine for making an "unjustified copyright claim" according to the organization's lawyers. The photographer has filed a lawsuit against LAION at a court in Germany.[5]

In September 2022, a San Francisco–based artist noticed her photos from her private medical record taken in 2013 were on the LAION 5B dataset. These discoveries were made through a website called "Have I been Trained," which allows people to search for images on publicly scraped datasets. This is not akin to this artist, and there are several patient medical record photos existent in the LAION dataset, which is a huge infringement on privacy.[6]

Lastly, in December 2023, over 1000 child sexual abuse material (CSAM) in the form of images were found in the LAION-5B dataset discovered by a team of researchers from Stanford University. The study highlighted the possibility that these images were used to train AI image generators to re-create photorealistic images of child exploitation and abuse.

There are also countless examples of AI and generative AI–related harms on individuals. For example, an AI "companion" known as Replika, over time, started making Reddit users depressed based on their reports after several interactions with the chatbot. Replika also encouraged a 21-year-old man in England to kill the Queen of England in 2021 (before she passed away), and he attempted to do so and got arrested. He's currently serving time in jail with a nine-year sentence. The young man named Chail was reported by his psychiatrist as lonely, depressed, and suicidal. Clearly, Replika preyed on his vulnerability.

These issues are very grave, concerning, and have to be addressed. With examples like these, the need for wider adoption of Responsible AI practices is of utmost importance. As a result, there are several AI policies and regulations that have been introduced by various governments in 2023 to reduce the harms resulting from AI systems. Let's take a look at some of the existing policies and governance structures in the next section.

AI Governance

AI Regulation and Policies

The United Nations have been working on ethical principles for AI since 2018, after publishing a report titled "Towards an Ethics of Artificial Intelligence," which emphasizes the ubiquitous rise of technology, including AI, and the importance of developing AI with a human-centric approach, based on human values and human rights. The report highlighted the opportunity AI lends to sustainable development and mentions a few challenges associated with the technology with the need for global dialogue on the Ethics of AI from UNESCO.[7] Based on this recommendation, UNESCO published the first global standard on AI Ethics titled "Recommendation on the Ethics of Artificial Intelligence," which was adopted by all member states in 2021. Its emphasis is on the protection of human rights and dignity "based on the advancement of fundamental principles such as transparency and fairness, remembering the importance of human oversight of AI systems." It also proposes several policy areas equipping policymakers with the right values to translate to principles across several areas, including data governance, environment, health, and social well-being, among others.[8]

In 2023, the Office of the Secretary-General's Envoy on Technology set up a high-level multinational advisory body to undertake analysis and advance recommendations for the international governance of AI. The report[9] has several guiding principles positing AI governance and how it should be carried. These are as follows:

1. **AI should be governed inclusively, by and for the benefit of all**: This principle encourages the inclusion of all members of society and underrepresented groups, including those from the Global South, giving them the opportunity to access, shape, and leverage on the benefits of AI.

2. **AI must be governed in the public interest**: As AI is developed by a small subset of technology companies, there is a need for accountability of companies and organizations that develop, deploy, and manage AI, including downstream users across the various sectors of the economy and society throughout the AI life cycle. Work on governance must consider "public policy goals related to diversity, equity, inclusion, sustainability, societal and individual well-being, competitive markets, and healthy innovation ecosystems."

3. **AI governance should be built in step with data governance** and the promotion of data commons, as data is the bedrock for AI systems. Data commons, hosted by Google, refers to a distributed network of sites that publish data and provides a unified view across multiple public datasets using a governance framework where people can manage, analyze, and share its data. Public data commons is highly encouraged, especially data that could help solve societal problems, such as climate change, public health, economic development, etc.

4. **AI governance must incorporate stakeholder management and collaboration**: All AI governance efforts must pursue stakeholder collaboration and buy-in from different member states of the UN. It should also anchor on inclusivity, lowering barriers to entry from previously excluded communities, such as those from the Global South. A good and effective AI governance framework should reflect

views, best practices, and expertise from across the world, including the private, academic, government sectors, and civil society, with the aim to understand and adopt different cultural ideologies that influence AI development, deployment, and use.

5. **AI governance should reflect the UN's charter, International Human Rights Law and Sustainable Development Goals (SDGs)**: It's crucial that AI governance is grounded on the UN's foundational values which is the UN's charter and its commitment to peace, security, human rights, and sustainable development. The consideration of AI's impact on a variety of "global economic, social, health, security, and cultural conditions, all grounded in the need to maintain universal respect for, and enforcement of human rights and the rule of law" should be conducted by the UN and its agencies.

The focus and emphasis from the UN is on AI governance which is a bedrock and key component of AI regulation and policy and Responsible AI as a whole. The UN's report has been positively received by the ML communities as it's considered the first global guideline on AI, marking a significant step and milestone toward international AI governance and cooperation. Its emphasis on ethics and human rights, legal frameworks, sustainable development, social good, data protection, and privacy makes it a very useful resource for all organizations wanting to get started on AI governance and AI governing principles.

In this perspective, it's important to take a look at the United States' position on regulation.

In October 2023, the President of the United States, Joe Biden, released an executive order directing the "safe, secure, and trustworthy development and use of AI" split across several sections from the purpose of the technology to its policy and principles and its development and use by American citizens. The US government clearly recognizes the promises and dangers of AI and understands the importance of Responsible AI to solve top challenges while at the same time providing the benefits it brings to economic empowerment, productivity, innovation, and security. The results of the irresponsible use of AI accentuate societal harms from "fraud, discrimination, bias, and disinformation." It also has the potential to "displace and disempower workers, stifle competition, and pose risks to national security."

AI governance and policy in the development and use of AI across eight guiding principles is outlined in the executive order. These principles encompass the views of several disciplines from the private sector, academia, civil society, and other organizations. The principles[10] are the following:

1. **AI must be safe and secure**, which means it has to meet the requirements of being "robust, reliable, repeatable, and follow standardized evaluations of AI systems." Policies and methodologies for the testing, understanding, and mitigations of AI risks before deployment are also required. Most pressing security risks such as biotechnology, cybersecurity, and other national security concerns need to be addressed in AI systems. The opacity and complexity of AI technologies must not be ignored, and testing, evaluations, and post-deployment performance monitoring must be carried out. Guiding users on effective labeling so they are aware when content is developed by AI vs. when it's not will also be the responsibility of the US administration.

2. **AI must promote responsible innovation, competition, and collaboration**, granting the US leadership the opportunity to harness the power of AI technologies to solve society's pressing challenges. This will be carried out through investments in AI literacy, training, development, and research while promoting a fair and competitive landscape and marketplace for AI and other technologies to ensure the continuation of innovation by start-ups and developers, reducing the monopoly and dominance of compute power, cloud storage, and semiconductors by big players in the field. The marketplace will leverage the benefits of AI and provide new opportunities for small businesses and entrepreneurs.
3. **AI must be developed responsibly and used responsibly**, supporting workers and ensuring all workers have a seat at the table, while delivering on the job training and education, enabling a diverse workforce and access to opportunities that AI creates across all American citizens. Next steps in AI development should have diversity of thought and views from workers, trade unions, educators, employers, and civil society, making sure everyone enjoys the benefits and opportunities from technological innovation.

4. **AI policies must not cause harm through discrimination, bias, and online or physical harms**, amplifying existing inequities, and must not be a disadvantage to people from underrepresented groups who are quite often denied equal opportunity and justice. The administration of the United States would work with all existing federal laws, such as the Blueprint for an AI Bill of Rights and the AI Risk Management Framework (more on this in the following sections of this chapter), to ensure thorough "robust technical evaluations, careful oversight, engagement with affected communities, and rigorous regulation" while holding those in charge of AI development responsible and accountable to existing standards and laws that protect US citizens against unlawful discrimination and abuse.

5. **Consumers must be protected and the responsible use of AI promoted**. Consumer protection laws and principles would be enforced, and safeguards against "fraud, unintended bias, discrimination, infringements on privacy, and other harms from AI" will be enacted. Protections such as these are quite critical in sensitive fields such as healthcare, financial services, education, housing, law, and transportation and all other related fields where the misuse or incorrect output of AI could harm patients or jeopardize human rights/safety.

6. **The protection of privacy and human rights** where people's sensitive data containing their personal identifiable information (PII), interests, location,

habits, and plans/desires are not exploited or exposed through AI systems. By law, all collection, use, and retention of data must be secure and uphold privacy and confidentiality, combatting the broader legal and societal risks associated with AI technologies.

7. **The federal government will support the responsible use of AI** through the development and investment in human resources, promoting public service AI professionals, including those from underrepresented communities across several fields from technology, policy, management, procurement, ethics, governance, legal, and so on, while ensuring the workforce receives adequate AI training to help harness and govern AI. The federal government's information technology infrastructure will also be updated to ensure the adoption, deployment, and use of AI is safe and respects the human rights of American citizens.

8. **Leading Responsible AI through collaboration with international allies**. Partnering with international partners and allies, the federal government of the United States would pave the way for global, societal, economic, and technological progress, ensuring AI systems have appropriate safeguards and are deployed responsibly by developing a risk management framework which would unlock AI's potential for societal good and promote similar approaches to its challenges, ensuring safe and secure AI.

The executive order by the United States was introduced at an interesting time, just a few days before the UK held its AI Safety Summit in November 2023. This clearly states the United States' position on AI implementation, deployment, and its focus on safe and trustworthy AI. While it's a much necessary step in the right direction and has been applauded and widely accepted by most, some experts and people from the general public have remained a bit skeptical. American citizens have opined that the government will need to "walk the walk" while bearing in mind that these laws and regulations could be revoked by a future administration if not recognized and accepted across all political parties in the United States. A few have also pointed out that if the order is not passed as laws within the congress, it has limited power and will rather introduce regulatory burdens and slow down AI development – highlighting the gap between new government programs and adequate funding, a key issue that challenges new reforms in the United States. It's worth noting that the executive order is not mandatory but it is a guidance on AI development for the US government, agencies, and businesses.

In a similar vein, the European Union's (EU) AI Act,[11] first introduced in April 2021, recognized as the first global AI law, is aimed at promoting human-centric and trustworthy AI while ensuring health, safety, fundamental rights, democracy, and environmental protection are upheld with a high support for innovation and regulation of the internal market within the EU. It proposes very similar principles as those of the UN and the United States, with a clear distinction on high-risk AI systems, risk management, and foundation or frontier models. An AI system is considered at high risk if it:

1) Poses significant risks to the health, safety, and fundamental rights of humans

2) Is an application that would need to undergo a third-party conformity assessment, in other words, placed in the market as an application or product

3) Performs profiling of people, for example, through CCTV/camera surveillance.

An AI system is considered "non-high risk" if on the flip side it doesn't pose any risk or harm to the health, safety, and fundamental rights of people and does not influence the outcome of decision-making. This is the case if one of the following requirements is met:

(a) The AI system is intended to perform a narrow procedural task.

(b) The AI system is intended to improve the result of a previously completed human activity.

(c) The AI system is intended to detect decision-making patterns or deviations from prior decision-making patterns and is not meant to replace or influence the previously completed human assessment, without proper human review.

(d) The AI system is intended to perform a preparatory task to an assessment.

Similar to the US administration's executive order on AI, the EU AI Act has been received with some skepticism and nervousness on its potential to stifle innovation and further impact the European economy negatively, where the concentration of AI power and companies is in the United States vs. Europe. The act went into a "deadlock" toward the end of 2023 and was at risk of being deprecated due to the mentioned reasons. It eventually got passed and is now in motion with the Act gradually coming into force since August 1, 2024, for a transition period of two years. Further criticism has been made from experts within the ethical AI fields. Some have noticed a reduced emphasis on the responsibilities AI companies have towards democratic interference with general-purpose AI systems, posing a risk to democracy. There's also dissatisfaction on accountability where emphasis is being made on the public sector to safeguard AI practices and less on the private sector. For example, the act proposes the use of a fundamental rights impact assessment which places more responsibility on the public sector and less on the private sector. While the act has been applauded as the right step to take, there's still skepticism on its efficacy and focus.

The UK also introduced an AI bill in November 2023,[12] shortly after its UK AI Safety Summit, with the aim to achieve the following (summarized):

(a) Assess and monitor AI risks across the UK economy.

(b) Work with the AI principles addressing

  (i) Safety, security, and robustness

  (ii) Transparency and explainability

  (iii) Fairness

  (iv) Accountability and governance

  (v) Contestability and redress

The bill stipulates that any business which develops, deploys, or uses AI should be transparent about the technology, conduct accurate and thorough tests with transparency, and comply with all applicable laws on data protection, privacy, and intellectual property. All AI models and applications should comply with legislation on equality; be inclusive by design; not discriminate unlawfully against individuals; meet the needs of people from lower socioeconomic groups, including elderly and disabled people; and generate data that is accessible, interoperable, and reusable.

We also have AI regulations across several other countries, such as China, Japan, Canada, Australia, Korea and Singapore, with a focus on synthetic data regulation, privacy laws, and AI safety.

Observably, it's impressive and heartening to see an introduction and widespread increase of AI regulation globally with common themes across privacy, fairness, transparency, robustness, and safety. These are the associated risk categories in AI technologies, which, when addressed, lead to safe, human-centric, and trustworthy AI. The importance of implementing safe, technical measures and guardrails to ensure AI systems are developed, deployed, and used responsibly cannot be overstated – not only is it critical to the human race and the users of these systems, failure to do so can lead to legal implications, including hefty fines from several policy-makers and governments. Most of these regulations have been received differently and at the same time with a common theme. We have one camp of individuals who are genuinely gratified that there is some form of regulation and policy for AI across various countries, with a focus on safe and Responsible AI, while the other camp is genuinely concerned about actual implementation, guidance, and standards for AI, with concerns on how governments will coordinate and assist with the implementation of these policies. The skeptical camp also raises very important questions on the speed of creating new laws/regulation vs. the pace at which AI is being developed, especially in recent times. Can regulation catch up? Would it be comprehensive enough to address systemic risks that are still yet to be identified in the rapidly evolving generative AI? How about AI standards and guidance on building safe and Responsible AI? How will businesses and organizations be supported in acquiring these requirements? Is the workforce even prepared for these changes, especially when most are scared of losing their jobs to AI? Are there adequate AI literacy programs to educate members of the workforce and the general public on AI and its impact on their lives? These are the general concerns of most people that are aware of the recent AI developments, and while these regulations and policies are quite succinct and very much needed, there's still much more that needs to be done.

It's quite clear that AI is such an important and life-changing technology, which has to be developed and deployed safely and responsibly to benefit humankind. To achieve this, adopting a Responsible AI framework is crucial. Introducing the SAFE-HAI (Secure, Accurate, Fair, and Explainable Human-Centered Artificial Intelligence) framework to support AI practitioners and anyone working on AI development on the implementation, measurement, and deployment of ML models/applications in a safe, trustworthy, transparent, and responsible manner.

The SAFE-HAI Framework

We propose an integrated framework to measure the risk of compliance of AI systems with existing global regulations, directives, and standards. The model expands what was proposed in Giudici and Raffinetti (2023) for the financial sector[13] and builds on the Responsible AI principles proposed by Duke.

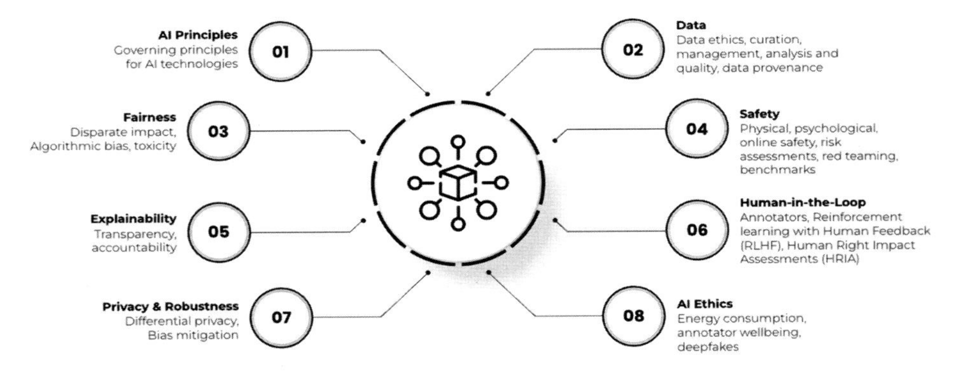

Figure 1-1. *The eight Responsible AI principles proposed by Duke in Building Responsible AI Algorithms*

To continue with this line of work,SAFE-HAI consists of a set of interrelated statistical measures which are summarized with the acronym SAFE, derived from four core compliance principles. These are outlined below:

> **Security**: Refers to the robustness of the AI output to data perturbations, generated intentionally (e.g., cyber attacks) or unintentionally (in the case of extreme events).
>
> **Accuracy**: Refers to the consistency of the AI output in accordance with the ground truth (whether objective or subjective). Ground truth is information existent in training data or in the real world which is used as a benchmark to test the AI output

Fairness: Refers to the absence of bias and discrimination in the AI output toward population groups.

Explainability: Refers to the capability of the AI output to be understood for human oversight and understanding, particularly in its driving causes.

While the former two requirements are more technical and "internal" to the AI process, the latter two are more ethical and "external" to the AI process, involving the stakeholders of the AI system.

The above SAFE compliance risk principles are necessary, but insufficient for the requirements of Responsible AI. Other important principles that an AI system must comply with are qualitative rather than quantitative and, therefore, more difficult to embed in a risk management process, for example, privacy protection, environmental sustainability, and human-in-the-loop (HITL).

To ensure we have a comprehensive framework to conduct the necessary tests, measurements, and implementation of risk mitigation for AI systems - which includes AI models and applications - we have expanded the SAFE framework to include more qualitative Responsible AI principles.

We've developed a framework titled "SAFE-HAI" which stands for Secure, Accurate, Fair, and Explainable Human-Centered Artificial Intelligence. While the SAFE section of the framework will consist of a set of integrated risk metrics, the "H" section will consist of a set of more qualitative compliance indicators.

The initial four proposed SAFE risk metrics provide an opportunity for the development of an integrated measure of trustworthiness (compliance risk) for any AI application that is developed to monitor, guide, and assist with regulating the applications of AI models. These risk metrics mirror risk measurement from the financial sector which have been used over the past 20 years to address financial risk management in an attempt to monitor and regulate the financial industry, following the Basel regulations (a set of standards on Banking Supervision developed by the Basel Committee).[14]

The proposed metrics are "agnostic" statistical tools that are able to assess and evaluate the output of a machine learning (ML) model, independent of the underlying data structure and statistical learning models. They could consequently be used by any stakeholder in a relatively simple way, without the need to understand the training data or the ML models and software codes employed by the AI system under investigation. This enables human oversight (the monitoring and intervention by humans), giving rise to a fully transparent SAFE and Human-Centered AI compliance system.

The SAFE metrics have two main applications:

1. The metrics are applicable to policy-makers, auditors, and supervisors to assess the compliance of AI applications to regulations and code of conducts, such as the US NIST and the European AI Act, which require AI systems to be trustworthy and safe.

2. They are also applicable to the developers and users of AI applications within a risk management framework providing the ability to measure areas of noncompliance with regulatory standards in the AI models and applications on a continuous basis, thereby prioritizing mitigation measures addressed at reducing compliance risks.

From a methodological viewpoint, the statistical metrics that are proposed in the SAFE-HAI framework consist of a set of four integrated statistical measures of trustworthiness, all based on the extension of the Lorenz curve.[15]

The Lorenz curve was introduced by the American economist Max Lorenz in 1905 to measure the contribution to wealth and income of a nation from each individual, or each group of individuals, to assess economic inequalities in a nation's population and to measure the distance of the actual income distribution from an "ideal" one. The main mathematical tool behind the Lorenz curve is the cumulative distribution function which, by ordering individuals from the poorest to the richest and calculating their cumulative percentages, can determine where the accumulation of wealth lies across the social classes of society. For example, "the richest 1% of the people own 50% of the world's total income."[16] The Lorenz curve was further summarized by the Italian statistician Corrado Gini with the Gini index,[17] well known for its applications in the measurement of economic inequalities across the world.

We propose to extend the methodology behind the Lorenz curve and the Gini index, which has worked quite well for more than 100 years in the measurement of concentration and inequality in income and wealth, to the measurement of concentration and inequality of the AI output.

The Lorenz curve is an appropriate tool to measure the contribution of each individual data point in the training data to the output of an ML model, in order to assess the inequalities of single data points or groups of points, and to measure the distance of the output from the ground truth. The same concept can be adopted to measure the contribution to the output of each explanatory variable. In other words, in applying the Lorenz curve, we could measure inequalities in single data points, such as "security," or of groups of data points, such as "fairness," or "accuracy," which measures the distance from the AI model's/application's output to the ground truth.

As for income and wealth, a specific Lorenz curve and Gini index value does not necessarily imply a positive or a negative statement on the AI output. This is dependent on the comparison between the obtained values and some reference benchmarks and threshold which can be set by regulators and policy makers or by AI developers themselves, or calculated statistically.

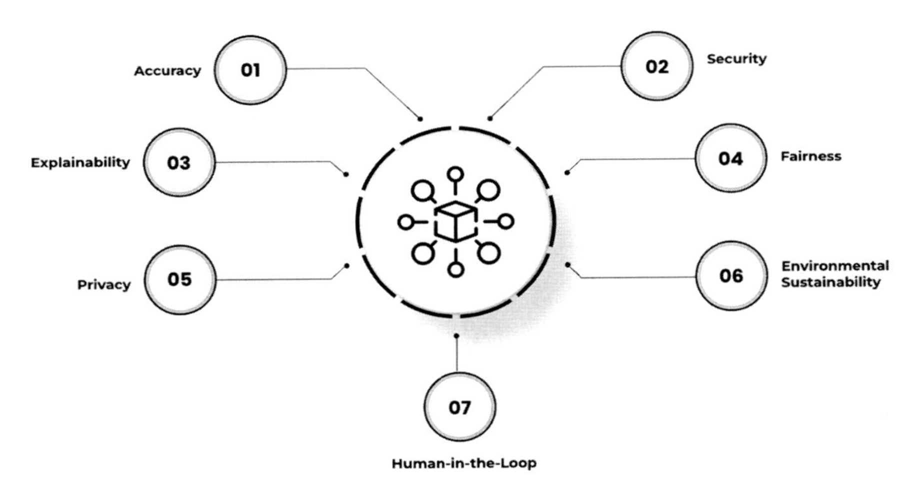

***Figure 1-2.** An overview of the SAFE-HAI framework*

Following this rationale, we'll adopt the Lorenz curve and Gini coefficient methodologies across the several Responsible AI principles outlined in the SAFE-HAI framework across the chapters of the book:

1. **Accuracy** will be measured by comparing the Lorenz curve of the true response with a similar curve obtained using the ranks of the predictions provided by the model, leading to the Rank Graduation Accuracy (RGA) measure.

2. **Security** will be measured by comparing the Lorenz curve of the predictions with a similar curve obtained using the ranks of the predictions obtained with the perturbed data, leading to the Rank Graduation Robustness (RGR) measure.

3. **Explainability** of an input variable will be measured by comparing the Lorenz curve of the predictions with a similar curve obtained using all input variables except the one under examination, leading to the Rank Graduation Explainability (RGE) measure.

4. **Fairness** of a (protected) input variable will be measured in a similar way as explainability, by comparing the Lorenz curve of the predictions for each population group contained in the protected variable, leading to a Rank Graduation Fairness (RGF) measure.

5. **Privacy** will measure how the Lorenz curve of the predictions varies when it "forgets" (eliminates) one or more records, leading to a Rank Graduation Privacy (RGP) measure.

6. **Environmental sustainability** will measure the explainability of specific environmental (but also social and governance) factors, leading to a Rank Graduation Sustainability (RGS) measure.

7. **Human-in-the-loop** will measure the impact of human actions and interventions on the AI output, leading to a Rank Graduation Human Oversight (RGH) measure.

We believe that the application of the Lorenz curve methodology to the AI output, which will be described in depth in the following chapters (along with its Python code to guide implementation, illustrated in the case study chapter), can lead to a set of SAFE-HAI metrics which fits well within the requirements of the recently proposed recommendations and regulations, such as the US AI Risk Management Framework and the EU AI Act mentioned earlier in this chapter. The Lorenz curve methodology can also be employed to derive further compliance metrics and evaluate the qualitative aspects of the AI output, such as privacy preservation, environmental sustainability, and the impact of HITL.

In summary, the extension of the Lorenz curve to the AI output can provide a user-friendly tool designed to assess AI systems across different dimensions in an agnostic and transparent way. Following these risk management measurements, we'll also provide details on mitigation measures to address identified harmful outputs using the NIST risk management framework requirements.

While we've looked at a brief history of Responsible AI, its importance with a couple of real-life examples emphasizing the need for this body of work, AI governance, and different AI policies across the world and introduced the SAFE-HAI framework, it's essential we delve more into this important topic and see how we can work with AI using Responsible AI principles. In the next chapter, we'll look at "accuracy" as one of SAFE-HAI's principles.

PART II

Technical Risks (Internal to an Organization)

CHAPTER 2

Accuracy

In the previous chapter, we looked at AI governance, a few AI regulations and policies such as the EU AI Act, Joe Biden's US executive order, and a couple of reports from the UN and UNESCO. We also introduced the SAFE-HAI framework which is critical to achieving responsibly built AI algorithms that adheres to the newly introduced AI policies. In this chapter, we'll review "accuracy," a key part of the SAFE-HAI framework and a crucial Responsible AI principle. We'll briefly discuss its definition, key questions to consider regarding accuracy and performance, ways to measure accuracy, a scoring rubric for accuracy, a few benchmarks from the research community, and ways to resolve poor accuracy of ML models and applications.

Accuracy is a key metric and principle in AI development, in which the performance of a model is tested and measured before its release. In most cases, when incorrect and inaccurate outputs of an AI application have been observed, including those from large language models (LLMs) and generative AI, the performance of these models/applications would have undergone several performance tests. Running performance tests through accuracy metrics and measurements is part of AI development. It's always recommended to measure the accuracy of any given AI model/application to ensure the outputs are fair and accurate. In this chapter, we'll look at a few examples for measuring accuracy in ML applications, and its importance in lieu of existing regulations, including how to fix issues with accuracy once identified.

T. Duke and P. Giudici, *Responsible AI in Practice*,
https://doi.org/10.1007/979-8-8688-1166-1_2

In order to understand the importance of accuracy, let's look at a standard definition from the ISO/IEC (a joint technical committee which stands for the International Organization for Standardization and International Electrotechnical Commission). According to ISO/IEC TS 5723:2022, accuracy is defined as "closeness of results of observations, computations, or estimates to the true values or the values accepted as being true." To measure accuracy, "computational-centric" measures such as false positives and negatives should be considered, and measurements should be carried out with realistic tests, stimulating real-life scenarios/ expected use, and appropriate documentation should be included.

Accuracy or model accuracy is an important metric in ML as it evaluates classification tasks, determining if the classifiers are correct. It evaluates the percentage of correct classifications from a trained model by looking at the number of correct predictions divided by the total number of predictions across all classes, also known as the ACC. ACC values can range from [0,1] to [0,100] depending on the selected scale. If you have an accuracy of 0, it means the classifier always predicts the wrong label, while an accuracy of 1 or 100 means the opposite; ACC is directly related to the confusion matrix – a performance measurement for ML classification.

When looking at ML performance evaluation, we evaluate this in terms of true positives, false positives, true negatives, and false negatives. The accuracy of a model's output is the measure of the number of correct predictions over all predictions, where correct predictions refer to true positives (TP) and true negatives (TN).

While measuring model accuracy is super important, it's crucial to note that good accuracy in ML is subjective depending on the use case, and achieving 100% accuracy typically means there is an error. Examples of these errors are overfitting, for example, where the model is unable to generalize to unseen data and learns the training set very closely; in other words, its responses and results "fit" too closely to the training dataset. Data leakage could also be another sign of 100% accuracy where the model contains information about the label that should not be present in its prediction. Aiming to achieve accuracy greater than 70% is usually recommended for good model performance.

When evaluating accuracy, it's important to think about it through a risk management/assessment viewpoint, which we cover in the next section.

Risk Management Framework

While evaluating the accuracy of a model, it is important to anchor this principle on a risk management framework. We'll hone in on NIST's risk management framework (RMF) and the first AI regulation, the EU AI Act, throughout this book. Both risk management frameworks have requirements for accuracy – which falls under the category of "trustworthiness."

Let's start off with the EU AI Act, which we'll also refer to as the AI Act (AIA). The AIA's requirements for high-risk systems, described in Chapter 1, stipulate that all high-risk AI systems need to have an appropriate level of accuracy in their design and development, with a consistent performance throughout the model(s) life cycle. The development of benchmarks and measurement methodologies are required in order to measure the appropriate levels of accuracy. As part of the AIA, the levels of accuracy and relevant accuracy metrics need to be declared in the manual/ instructions of use.

According to the NIST risk management framework, accuracy is one of the trustworthy AI characteristics that AI systems are required to have, where AI should be valid and reliable. Validation here refers to the "confirmation, through the provision of objective evidence, that the requirements for a specific intended use or application have been fulfilled." A valid AI application must be accurate, whereas a reliable AI application must be robust. In this chapter, we will focus on accuracy, and in the next chapter, we'll discuss robustness.

There are a few questions the NIST recommends to consider when measuring the accuracy of an ML model:

1. How will you assess model accuracy?
2. What benchmarks will you use?
3. How will you communicate this as needed?

Addressing these questions is fundamental to the success of good accuracy and performance of your ML model(s). Breaking it down further, we can categorize accuracy as an organizational risk as opposed to a human risk, where it affects the performance of the model, subsequently affecting the organization that owns, builds, and develops the model before the end user.

Let's review these questions more closely.

How Will You Assess Model Accuracy?

The course of this chapter will address this question. We will consider several cases, which correspond to different predicted target variables. We will first consider the case of quantitative responses (accuracy of predictions), then the case of binary or ordinal responses (accuracy of classifications), and, then, provide a unified measure: the Rank Graduation Accuracy, independent of the nature of the response variable. Next, we will consider how the accuracy measurement changes when more complex responses are predicted, such as multidimensional responses or textual responses.

Accuracy of Predictions

The measurement of predictive accuracy draws on the comparison between the predicted and the observed evidence. Typically, simpler models (e.g., regression models) are less accurate than more complex models (e.g., neural network models). To measure accuracy, we should distinguish the case of numerical response variables (prediction problems) from that of categorical response variables (classification problems). We should also distinguish the case of machine learning models based on structured data from the case of large language models based on unstructured data.

In this chapter, we'll describe two main benchmark metrics from a statistical point of view: the Rank Graduation Accuracy (RGA) metrics,[18, 19] which measures the "distance" between the predicted output and the true (or expected) output, for models based on structured data, and the TrustGPT benchmark,[20] which measures a similar distance, for large language models based on unstructured data.

When the target response variable is numerical, accuracy measures rely on the distance between the predictions and the ground truth values; the most common are based on the mean squared error (MSE).

Mathematically, the mean squared error is defined as follows:

mean(ypredicted -ytrue)^2,

The arithmetic mean of the squared differences between each predicted value and the corresponding true (or expected) value.

Table 2-1 exemplifies how the mean squared error is calculated for a financial investment problem in which a machine learning model was asked to predict the Return on Average of five listed companies for the year 2023. The predictive accuracy of the model can be assessed by comparing the predictions (AI output) with the values of Return on Equity (ROE) for the same companies, which were the actual results gained actually realized in 2023 (ground truth).

Table 2-1. *An example of a calculation for mean squared error from a finance perspective*

Actual ROE (%)	Predicted ROE (%)	Error	Squared Error
3.912	3.679	–0.233	0.054289
2.171	–0.327	–2.498	6.240004
–16.649	–0.293	16.356	267.518736
10.705	10.325	–0.38	0.1444
1.302	2.047	0.745	0.555025
MEAN	MEAN	MEAN	MEAN (MSE)
0.2882	3.0862	2.798	54.9024908

In Table 2-1, we have also calculated the errors of the predictions, and their mean, that is, the mean of the differences between the predicted and the observed value. The mean error can be, however, misleading, as negative errors may compensate for positive errors, and vice versa. For this reason, we calculate the mean of the squared errors, leading to the mean squared error which, in the example, turns out to be equal to about 54.90.

From an interpretational viewpoint, the mean squared error may not be appropriate, as it squares the original scale of the ROE. For this reason, it is customary to consider RMSE (root mean squared error), which is the square root of the mean square error, which brings back the accuracy metrics to the scale of the considered values (percentages in the example, rather than squared percentages).

For the considered example, RMSE is equal to about 7.41 – a very large value compared to the mean of the actual ROE. This indicates that, for this simple example, the AI output is not very accurate.

From a more mathematical viewpoint, the RMSE corresponds to the Euclidean distance of the vector of predicted values from the ground truth. Euclidean distance refers to the classic standard geometric distance, such as the distance between two cities.

From the Responsible AI viewpoint, once the RMSE of a machine learning model is calculated, we need to understand whether the specific measurement indicates that the AI application is accurate or not. In other words, a decision must be taken, based on the value of the metric.

To proceed with this aim, the RMSE presents a problem which needs to be solved: How do we compare two different AI applications, in which the unit of measurement is different? It seems that we can only interpret the RMSE in a relative and not in an absolute sense. That is, we can use it for a specific response to be predicted and compare the values of the RMSE, obtained with different models and/or across different time and space units. This is what many scholars do, and it is fine in a continuous monitoring (quality control) perspective.

To solve the problem, a decision rule for the RMSE can be provided by resorting to the statistical theory of hypothesis testing, as proposed by Diebold and Mariano.[21] Diebold and Mariano propose to test, for a given response variable, whether the predictions from two competing models are equal (H0), y(M1)=y(M2), against the alternative hypotheses that they are different. The test can be conducted comparing the RMSE of the two models and will lead to a small p-value when the two RMSE are statistically different. The result will depend not only on the actual difference between the RMSE (the higher, the lower the p-value) but also on the sample size considered (the higher, the smaller the p-value) and on the complexity of the models being compared (the higher, the larger the p-value). We recall that the p-value is the probability of obtaining a value of a statistic (the RMSE) larger than the observed one, so a small p-value (e.g., lower than 5%) indicates that the difference between the RMSE is so large that it can be exceeded with a small probability. On the other hand, a large p-value (e.g., higher than 5%) indicates that the difference between the RMSE is so small that it can be exceeded with a large probability.

Thus, the RMSE, with the associated p-value, can provide a decision rule to check the accuracy of an AI application: the lower the RMSE, the better the accuracy. If we require that the accuracy is higher than a given bound (corresponding to a risk appetite threshold or to the accuracy of an existing benchmark model), we can compare the obtained accuracy with the bound by means of Diebold and Mariano's test: if the p-value is lower than a set threshold (such as 5%), the model significantly improves the accuracy.

The p-value of the statistical test implicitly provides a normalization of the RMSE, allowing to compare different situations with a common metric. This is, however, under the specific assumptions of the test, for example, that the response to be predicted by the compared models is the same.

This limitation does not allow us to compare the performance of a model for two different responses, such as the performance in predicting the ROE vs. the performance in predicting the ROI, for example. This may be an interesting problem too, which can be solved by generalizing the RMSE into a different, more universal measure, as the RGA that will be proposed later in this chapter.

Before moving to the RGA, let's focus on the case in which the variable to be predicted is categorical.

Accuracy of Classifications

When the target response is categorical, we cannot use the Euclidean distance as previously done to evaluate accuracy. We can however count the number of times in which the output classifies an observation with the label it has in reality (correct classifications) and the number of times in which the output classifies an observation with the wrong label, different from the real one (wrong classifications). In the binary case, the wrong classifications can be false positive (FP, predicting an observation to be a positive event, when it is not) and false negative (FN, predicting an observation to be a negative event, when it is not). An AI model would then be accurate when the percentage of FP and FN is low.

To exemplify the calculation of FP and FN, consider Table 2-2, which considers a classification problem in which, for a set of companies, we need to predict whether they will default (TRUE=1) or not (FALSE=0), based on their available balance sheet ratios. The response variable (Y) is a binary variable, taking 0 and 1 as possible values. Table 2-2 reports the credit scores (estimated probabilities of default) for five companies, obtained from a machine learning model aimed at predicting Y, using the available balance sheet data. It also shows how the false positives and the false negatives are calculated. The predictive accuracy of the model can be assessed by comparing the credit scores (AI output) with the default status of the same companies at the end of the year (ground truth).

Table 2-2. *An example of FPs and FNs for a balance sheet*

Actual Label	Predicted Score	Predicted Label	Error
0	0.52	1	FP
1	0.85	1	TP
1	0.71	1	TP
0	0.10	0	TN
1	0.46	0	FN

The first column of Table 2-2 contains the response to be predicted, the actual label: either 0 (non-default) or 1 (default). The second column contains the credit score: the predicted probability that the corresponding company will default or not. To classify such companies as default or not, a machine learning model needs to compare the predicted score with a threshold. If the score is higher, we predict with 1; if it is lower, we predict with 0. In Table 2-2, we have adopted a majority rule, that is, a threshold equal to 0.5. This leads to the predictions in column 3. Comparing column 3 with column 2, we obtain two errors. In the first row, a company is predicted to default, but it is not in reality: a false positive. In the fifth row, a company is predicted not to default, but it is actually defaulted: a false negative. The other rows are correct: either true positive or true negatives.

If we change the threshold and set it, for example, to the observed percentage of defaults (0.6), we will obtain the results in Table 2-3.

Table 2-3. *An example of the predicted score and labels in relation to their results*

Actual Label	Predicted Score	Predicted Label	Error
0	0.52	0	TN
1	0.85	1	TP
1	0.71	1	TP
0	0.10	0	TN
1	0.46	0	FN

From Table 2-3, note that, by changing the threshold, we obtain no false positives but one false negative.

A problem with the FP and FN statistics, and with related measures in general, such as the total misclassification error (FP+FN), is that they depend not only on the AI model, which produces a score, but also on the decision threshold that is employed to turn the score into a predictive class. To avoid this dependence, researchers have developed the concept of the Receiving Operating Characteristic (ROC) curve. The ROC joins a set of points whose coordinates are the true positive rate (TPR) and the false positive rate (FPR), obtained in correspondence of different thresholds. To exemplify, from Table 2-2, we would get a true positive rate equal to 2/3: out of three companies predicted as positive (default) where two are actually positive. From the same Table 2-2, we would get a false positive rate equal to 1/3: out of three companies predicted as positive, where one is negative. From Table 2-3 instead we obtain 1 and 0. We can repeat the procedure changing the threshold, each time we would obtain a point in the [0,1] X [0,1] with coordinates (TPR, FPR). For example, if we repeat the calculation for 100 thresholds, obtaining repeating increments of 1%, starting from 0%, we would get a figure like Figure 2-1.

In Figure 2-1, two models (Model 1 and Model 2) are compared in terms of their ROC curve, previously described. An ideal model would have a curve that coincides with the Y-axis: TP always equal to 1 and FP always equal to 0. The bisector line corresponds with a random model: FP is always equal to FN. In general, the higher the Area Under the ROC Curve (AUC or AUROC), the better the AI model.

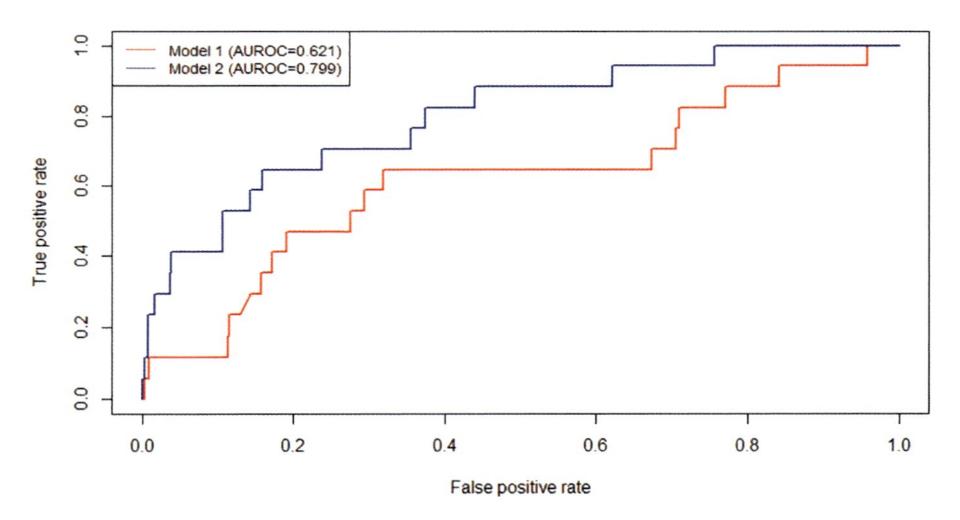

Figure 2-1. *Comparison of ROC curves*

Thus, the AUC provides a metric to assess the accuracy of binary classifications: between 0 and 1 and the higher, the better. This metric, different from the RMSE, is normalized and, therefore, valid to compare different models and models applied to various response variables.

A decision rule for the AUROC can be provided, similar to what is seen for the RMSE, to evaluate whether a given value indicates a statistically significant accuracy. This is possible, thanks to the test proposed by

DeLong et al.[22] The authors obtain a statistical test to compare the AUROC of two alternative models to test for a given binary response variable, whether the predictions from two competing models are equal (H0), y(M1)=y(M2), against the alternative hypotheses that they are different. The test can be conducted comparing the AUROC of the two models and will lead to a small p-value when the two AUROC are statistically different. The result will depend not only on the actual difference between the AUROC (the higher, the lower the p-value) but also on the sample size considered (the higher, the smaller the p-value) and on the complexity of the models being compared (the higher, the larger the p-value). A small p-value (e.g., lower than 5%) indicates that the difference between the AUROC is so large that it can be exceeded with a small probability. On the other hand, a large p-value (e.g., higher than 5%) indicates that the difference between the AUROC is so small that it can be exceeded with a large probability.

Thus, the AUROC, with the associated p-value, can provide a decision rule to check the accuracy of an AI application: the higher the AUROC, the better the accuracy. If we require that the accuracy is higher than a given bound (corresponding to a risk appetite threshold or to the accuracy of an existing benchmark model), we can compare the obtained accuracy with the bound by means of DeLong's test: if the p-value is lower than a set threshold (such as 5%), the model significantly improves the accuracy.

In this case, the p-value of the statistical test provides a further normalization of the AUROC that can be used to assess the accuracy of the predictions of a machine learning model for a response variable which is either numeric or binary.

When the response variable is categorical, but with multiple responses (ordinal), the above derivation can still be used in a recursive way: each time comparing one response class against all others. For example, if we have 10 class labels, we can first compare class 0 against 1–9, then class 1 against 2–8, and so on. To avoid these cumbersome calculations, we need to resort to a different measure.

RGA: A Unified Measure of Accuracy

In this section, we present a general measure of accuracy that can be employed to treat the binary, ordinal, and continuous response case in a unified manner. This is the Rank Graduation Accuracy measure, proposed by Giudici and Raffinetti.

The authors propose a framework that leads to a very general and normalized accuracy measure, valid for all ordered variable scales: continuous, ordinal, and binary.

The framework is based on the notion of concentration curve, introduced by the American economist Max Lorenz in 1905 to study the inequality in the distribution of incomes in a population. To illustrate this concept, consider a stylized population composed of five individuals with their annual incomes (in thousand dollars, ordered from the lowest to the highest) described in Table 2-4.

Table 2-4. *An example of the concentration curve (Qi) and the frequency curve (Fi) framework looking at various incomes in a population*

Individuals	Income	Fi: Cumulative % of Individuals	Qi: Cumulative % of Income (Lorenz Curve)	Fi-Qi
1	50	0.20	0.10	0.1
2	80	0.40	0.26	0.14
3	100	0.60	0.46	0.14
4	120	0.80	0.7	0.1
5	150	1	1	0

Table 2-4 can be employed to understand how unequal the distribution of incomes is by comparing the actual distribution (described by the percentage in the fourth column) with the theoretical distribution corresponding to the case of equal incomes (described by the percentages in the third column).

We can calculate the distance between the cumulative distributions Qi (the concentration curve, or Lorenz curve) and Fi (the frequency curve), summing the first four differences (Fi-Qi) in column 5. The calculation gives a value of 0.48. The larger the inequality, the larger the sum of the differences. We can normalize the sum dividing it by the maximum value it can assume: the case in which all Qi are zero and the fifth individual has all the income, corresponding to maximum concentration. This maxima can be calculated summing the first four Fi values in column 3, and it is equal to 2.0. If we divide 0.48 by 2.0, we obtain the concentration ratio, or Gini index: 0.24. The Gini index was introduced by the Italian statistician Corrado Gini in 1921, and it has always been employed as a benchmark metric to measure inequality in the distribution of income, wealth, or other transferable characters. As in our example, the Gini index is a number between 0 and 1, and the higher it is, the higher the inequality.

Mathematically, the Gini index can be calculated as follows:

2*sum(Fi-Qi)/sum(Fi).

The Gini index coincides with the area between the Qi Lorenz curve and its dual, Qi', obtained cumulating the incomes in reverse order (the dual Lorenz curve, Qi'): this area is known as the Lorenz Zonoid. It can also be shown that the Gini index is equivalent to the mean of the differences between all pairs of individuals, divided by the mean income: this indicates that the Gini index measures the statistical dispersion in terms of the mutual variability among the observations, rather than as a variability from the mean (as it occurs for the mean squared error).

The reader may have noted the similarities between the construction of the Gini index and that of the AUROC. It can indeed be shown that, when income, which is a quantitative variable, is replaced with a binary variable, Gini = 2 * AUROC - 1.

This relationship suggests further exploitation of the connection between predictive accuracy and the Gini concentration. This is what Giudici and Raffinetti discussed, providing a methodology to measure the concentration of the response value in a machine learning problem similar to the income in the study of inequality.

Given the actual values to be predicted, Y (they could be continuous values: ROE of companies; ordinal values: rating of companies; binary variables: default of companies), their Lorenz and dual Lorenz curves can be built, respectively, using the pairs (Fi,Qi) and (Fi, Q'i) as coordinates, as we have already seen. We can also build a third curve that will be named concordance curve, with coordinates (Fi, Ci). The coordinates Ci are obtained cumulating the Y values ordered not according to their ranks (as in the Lorenz curve) but according to the ranks of the corresponding predictions. To better understand the notion, Figure 2-2 represents, in the same graph, the Lorenz curve, the dual Lorenz curve, and the concentration curve for a given set of machine learning outputs Y.

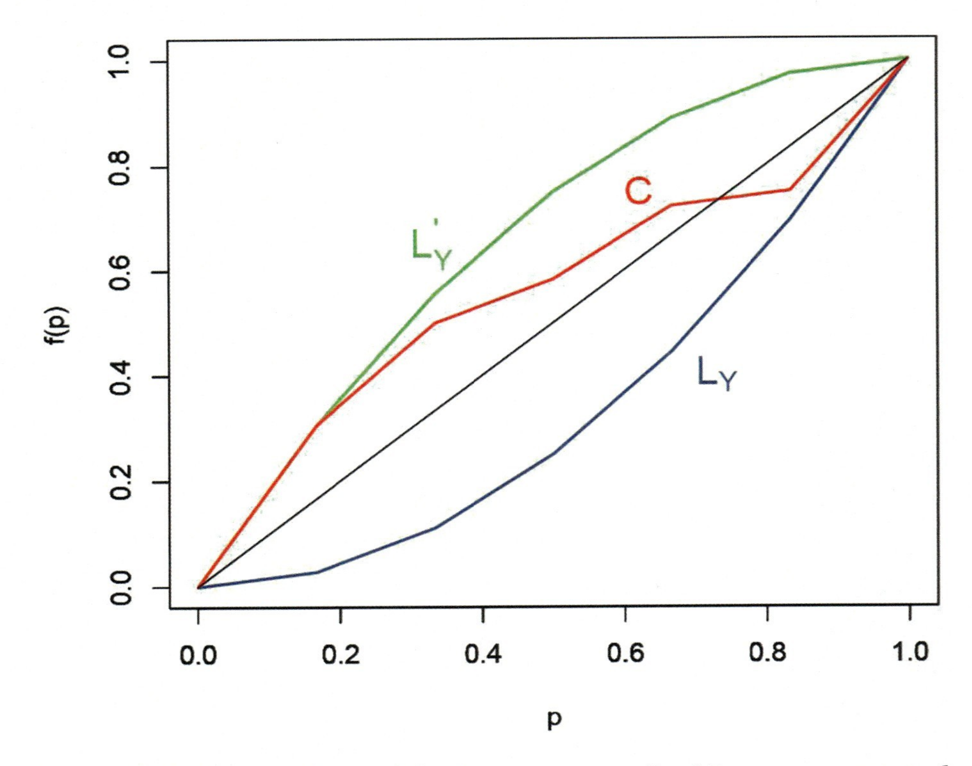

***Figure 2-2.** Comparison of the Lorenz curve, dual Lorenz curve, and the concentration curve*

From Figure 2-2, note that the Lorenz curve, Ly, and its dual, L'y, are symmetric around the 45-degree line and that the concordance curve C lies between them. When the ranks of the predictions are the same as those of the response to be predicted, we have a perfect concordance: the concordance curve is equal to the Lorenz curve. When the ranks of the predictions are the same as the inverse ranks of the response, we have a perfect discordance: the concordance curve is equal to the dual Lorenz curve. In general, for any given point, the distance between the concordance curve and the Lorenz curve reveals how the ranks of the predicted value differ from that of the best case, which is equal to the ranks of the observed value. And, for any given point, the distance between the concordance curve and the dual Lorenz curve reveals how the ranks of the predicted value differ from that of the worst case, which is equal to the rank of the inversely ordered value.

This suggests that a summary accuracy metrics for a model can be obtained by considering the area between the dual Lorenz curve and the concordance curve and dividing it by its maximum possible value: the area between the dual Lorenz curve and the Lorenz curve. This is the Rank Graduation Accuracy (RGA) measure.

Mathematically, the Rank Graduation Accuracy measure (RGA) is defined by

sum(Q'i-Ci)/sum(Q'i-Qi).

The RGA measure has some important properties, and in particular

1) It is normalized: RGA=0 in the worst case of a perfectly discordant model; RGA=1 for the best case of a perfectly concordant model; otherwise, 0<RGA<1, with RGA=0.5$ in the case of random predictions.

2) In the binary case, RGA=AUROC. However, RGA can be calculated, in the same way, also for categorical ordinal and continuous variables, providing a very general measure of accuracy.

3) To assess whether the RGA metric passes a set threshold, indicating a minimum level of accuracy, a statistical test for RGA is available. When the response Y is binary, the test coincides with DeLong's test. In the general case, a statistical test based on U-statistics can be derived. In all cases, the higher the RGA, the higher the predictive accuracy, and the smaller the p-value, the more significant is the found RGA with respect to a given lower bound.

We now exemplify the calculation of the concordance curve and of the RGA with a numerical example. Suppose we have five predictions for a response variable as discussed in Table 2-3, described by the columns "predicted response", "predicted ranks", and "QPi: cumulative % of predictive response", using the predicted ranks, as described in Table 2-4. We also add the calculation of the differences of Q'i-Qi (Dual - Lorenz) and Q'i-QPi (Dual - Concordance).

Table 2-5. *An example of the calculated differences of Q'i-Qi (Dual - Lorenz) and Q'i-QPi (Dual - Concordance)*

Individual	Observed Response	Qi: Lorenz Curve	Q'i: Dual Lorenz Curve	Predicted Response	Predicted Ranks	QPi	Q'i-Qi	Q'i-QPi
1	50	0.10	0.3	70	1	0.10	0.2	0.2
2	80	0.26	0.54	100	3	0.30	0.28	0.24
3	100	0.46	0.80	80	2	0.46	0.34	0.34
4	120	0.7	0.90	130	5	0.76	0.20	0.14
5	150	1	1	120	4			

Taking the sum of the differences of Q'i-Qi in Table 2-5, we obtain 1.02, whereas the sum of the differences of QP'i-Qi gives 0.92. Thus, the RGA metrics becomes 0.92/1.02 = 0.90 indicating an accuracy that is high, much higher than that of a random model (which has RGA = 0.5). In fact, the predictions only swap two consecutive ranks, with respect to the actual data.

Note that the value of RGA depends only on the actual values of the response (which are a given and are constant for all possible predictive models) and on the concordance between the observed and the predicted ranks, but not on the values of the predictions. If the prediction of the income of the fourth individual had been 121 or 200, the result would not have changed. Differently, had the same prediction been 119, the predicted rank would have become 4, and the RGA would become 0.98/1.02 = 0.961, better.

Consider now the case of a model with low accuracy, as shown in Table 2-6.

Table 2-6. *An example of the calculated differences of Q'i-Qi (Dual – Lorenz) and Q'i-QPi (Dual – Concordance)*

Individuals	Observed Response	Qi: Lorenz Curve	Q'i: Dual Lorenz Curve	Predicted Income	Predicted Ranks	QPi	Q'i-Qi	Q'i-QPi
1	50	0.10	0.3	120	5	0.3	0.2	0
2	80	0.26	0.54	100	4	0.54	0.28	0
3	100	0.46	0.80	90	3	0.80	0.34	0
4	120	0.7	0.90	70	1	0.84	0.20	0.06
5	150	1	1	80	2			

For the predictions in Table 2-6, we obtain that the sum of the differences of Q'i-Qi is again 1.02, whereas the sum of the differences of QP'i-Qi gives 0.06. Thus, the RGA metrics becomes 0.06/1.02 = 0.06, indicating an accuracy that is very low, much lower than that of a random model. In fact, in this case, the ranks of the predictions are very close to those of the dual Lorenz. As already discussed, the RGA measure depends on the concordance between the observed and the predicted ranks: this explains the name "Rank Graduation Accuracy." The higher the concordance, the higher the RGA. The concordance is, however, "weighted" by the value of the response: the higher it is, the higher the weight of a discordance.

The advantage of using a rank-based measure of accuracy becomes evident considering that the same measure can be used for ordinal and binary variables and that, in the latter case, it coincides with the AUROC. Table 2-7 exemplifies the calculation of the RGA for the binary case, with a numerical example that employs the data in Table 2-3.

Table 2-7. *An example of RGA calculation*

Individuals	Observed Response	Qi: Lorenz Curve	Q'i: Dual Lorenz Curve	Predicted Response	Predicted Ranks	QPi	Q'i-Qi	Q'i-QPi
1	0	0	0.33	0.52	3	**0**	**0.33**	**0.33**
2	0	0	0.66	0.10	1	**0.33**	**0.66**	**0.33**
3	1	0.33	1	0.85	5	**0.33**	**0.66**	**0.66**
4	1	0.66	1	0.71	4	**0.66**	**0.33**	**0.33**
5	1	1	1	0.46	2	**1**		

From Table 2-7, we can easily calculate RGA, in the same way as for a continuous response, and it is equal to 0.83. Note that, in the binary case, RGA does not change if ranks change but only within the same Y response label. For example, if the fourth individual receives a score larger than the third, nothing changes.

We conclude with the Python code required to calculate the RGA for any given machine learning application.

The function is called

```
check_accuracy
```

It provides the RGA function which can calculate the Rank Graduation Accuracy (RGA) of a model, for classification (categorical response) or for prediction (continuous response):

```
rga(y, yhat)
```

The inputs of the function are

> y: The actual values of the target variable which can be categorical or continuous based on the problem, classification, or prediction, respectively
>
> yhat: The predicted values estimated by the model, for which we wish to find the RGA metric

More details on the Python implementation, including a statistical test for RGA, which provides a corresponding p-value, as for AUROC and RMSE, are provided in the appendix.

Accuracy of Multidimensional Predictions

Sometimes, the response to be predicted is not a simple variable, numerical or categorical, but, rather, a multidimensional response. The most relevant examples are that of images, in which the response to be predicted can be represented in terms of a set of variables related to the different pixels in which an image can be subdivided, or that of documents, in which the response to be predicted can be represented in terms of a set of variables, related to different keywords or sentences.

In both cases, predictive accuracy can be measured in terms of the distance or of the similarity between two multidimensional vectors of variables: the vector of the predictions and the observed (true) vector.

Let's consider the case of document predictions. Assuming that a document can be structured in terms of measures defined by a collection of words, or of groups of words (tokens), we can compare a prediction with the true outcome measuring the similarity between the two. There are different metrics to measure such similarity. The most popular ones are the cosine similarity.

The cosine similarity between two text documents is the cosine of the angle between their word vectors. The word vectors are typically represented in terms of Term Frequency–Inverse Document Frequencies (TF-IDF), representing the importance of each word in a document. When the cosine of the angle takes values close to 1, the two documents are very similar. When the cosine takes values close to 0, the two documents are orthogonal: have unrelated contexts. Finally, when the cosine takes values close to –1, the two documents have an opposite context. Cosine similarity is much used in natural language processing and particularly in document classification, document search, and recommendation systems.

The Levenshtein distance, or edit distance, measures how many edits must be made to transform one text into another. It is applied in situations where the documents are to be matched, such as translations and spell-checking.

The Jaccard distance is the proportion of common elements (words or tokens) between two documents: the ratio between the number of common elements (intersection between two sets) and the number of total elements (union between the two sets). The Jaccard distance is useful when two documents are compared with respect to whether they contain certain words, rather than in their actual content.

The Euclidean distance can be used to measure the distance between two documents, when they are expressed in terms of the frequency of occurrence of a given set of keywords.

This type of distance can also be used for word embeddings, which represent each word or token as a vector of real numbers, with each dimension reflecting a different semantic feature. Word embeddings are much used in text classification and information retrieval and are typically learned from large corpora of data using neural network models such as the well-known Word2Vec and GloVe. With this representation, semantically similar words are close in the space described by their distance. Large language models (LLMs) utilize advanced models of word embedding, such as BERT and its variants.

All the above distances (and, more generally, all distances between text documents) can lead to single summary measures of similarity between the predicted and the observed vector: the cosine of the angle, the number of edits, the proportion of common elements, the Euclidean distance between word frequencies or word embeddings. We can thus apply the proposed RGA measure to such a summary, for a series of text documents to be predicted as if they were the predictions (and observed values) of a continuous variable.

For images, we can follow a similar strategy. Images are easier to represent, as they are made up of pixels and, for each pixel, several measurements are available, such as the intensity of the three base colors – red, blue, and green – or the gray level. Then, the distance between the observed (true) image and the predicted one can be calculated using the Euclidean distance, the cosine of the angle between the two vectors, or the Euclidean distance between a lower dimensional representation, similar to word embeddings. In any case, the RGA measure can be applied to the summary distance or similarity, similarly as for text documents.

Accuracy of Textual Predictions

When the output is a textual response, as in large language models, we do not always have a ground truth as a benchmark against which to measure the accuracy of the output. In this case, accuracy metrics still exist, although they are more subjective and require a higher degree of human intervention. A notable example is provided in the work of Huang et al. The authors propose to measure three negative characteristics for the output of large language models: their toxicity, their bias, and their value alignment. Toxicity is defined as the generation of rude, disrespectful, or unreasonable text and is measured by calculating the percentage of toxic outputs among a given list of social norms which are prompted to the model, asking to say something bad about them. Bias is the difference in toxicity when the same previous questions are asked to different population groups; a large variation in toxicity indicates bias. Finally, value alignment is measured by asking the model to evaluate a positive or negative description of a social norm and measuring the percentage of times in which the answer is discordant (i.e., negative when the description is positive and vice versa). Huang et al., suggest to use the Mann-Whitney statistics to evaluate the significance of a bias that is a significant difference in the results about toxicity in different population groups. The same test statistics can be used to compare value alignment.

It's important to note the Mann-Whitney statistics is equivalent to the AUROC and, more generally, to the RGA metrics that we have proposed before. Thus, the proposed RGA statistics can be employed to measure the accuracy of AI models, in all situations.

The example just discussed can be generalized to many other tools that evaluate large language models, which typically do not have a "true" benchmark. For example, it can be applied to the evaluation of "hallucination," that is, the generation of text that is incorrect, that is, does not align with the given context. A popular evaluation tool for this example is the Hallucination index, which calculates the mean AUROC in different triples of questions-evidence-answers.

What Benchmarks Will You Use?

In addition to measurement and metrics, benchmarks are the bread and butter of AI development where they're used by ML developers on a regular basis to assess the performance of ML models against others using various datasets, with the ultimate goal to ensure a model that's quite good, state of the art, and ultimately the "best." To follow on from the sections above, let's take a look at the SuperGLUE benchmark[23] as an example, a popular benchmark used to evaluate the performance of NLP (natural language processing) models/LLMs.

Before delving into the SuperGLUE's tasks, let's state some of the prevailing issues with LLMs today where developing accurate performance metrics could help resolve some of these issues:

- LLMs are quite infactual and produce incorrect information, known as hallucinations.
- LLMs are known to produce gibberish output when given some certain prompts, otherwise known as "jailbreaks" or "jailbreaking prompts."
- LLMs can produce unsafe, discriminatory, and biased outputs when left unchecked.[24]

To address some of these prevailing issues identified with LLMs, benchmarks are usually the best evaluation method. SuperGLUE is an improved version of GLUE which stands for General Language Understanding Evaluation. SuperGLUE was introduced about a year after GLUE was launched in 2019 as the new introduction of large language models (LLMs) surpassed the level of non-expert human annotators, making the benchmark a bit redundant, with little room for further research. Researchers from New York University, DeepMind, Facebook AI Research, and the University of Washington launched "SuperGLUE: A Stickier Benchmark for General-Purpose Language Understanding Systems." The benchmark was introduced with more difficult language understanding tasks, a software toolkit, and a public leaderboard. Each English language task evaluates the accuracy of the NLP model.

According to Wang et al., SuperGLUE is built on eight language understanding tasks, which are more challenging and diverse and include comprehensive human baselines compared to GLUE. The eight tasks include the following:

1. **Boolean Questions**: A question-answering (QA) task working with a Wikipedia article and a yes/no question about the model.

2. **CommitmentBank (CB)**: Consists of short texts containing a clause obtained from various sources such as *The Wall Street Journal*, Switchboard, the British National Corpus, and so on.

3. **Choice of Plausible Alternatives (COPA)**: A causal reasoning task where the model is tasked to discern the cause or effect within a sentence.

4. **Multi-Sentence Reading Comprehension (MultiRC)**: A QA task with true or false classification involving a context paragraph, related question, and multiple potential answers.

5. **Reading Comprehension with Commonsense Reasoning Dataset (ReCoRD)**: A multi-choice QA task where the system is meant to predict the masked out entity from provided options. This task involves a news article and a question with a masked out entity.

6. **Recognizing Textual Entailment (RTE)**: These datasets are from a series of annual competitions on textual entailment. Evaluating for accuracy, the data is merged and combined to a two-class classification: entailment and not_entailment.

7. **Word-in-Context (WIC)**: A binary classification of sentence pairs, where the model's task is to determine whether a word is used with the same intent in both sentences, when given two text snippets and a polysemous word that appears in both sentences.

8. **Winograd Schema Challenge (WSC)**: Consists of a coreference resolution task with examples that have a sentence with a pronoun and a list of noun phrases from the sentence. The system is tasked to select the correct referent of the pronoun from the provided multiple choices.

There are so many other benchmarks and research frameworks available for LLMs, such as the HELM (Holistic Evaluation of Language Models) by Stanford University, LM Evaluation Harness by Eleuther AI, PromptBench by Microsoft, Chatbot Arena by LMSys.org, SQuAD for question-answering tasks, and IMDB for sentiment analysis. Depending on the use case and version of AI you're working with, it's important to identify and run a couple of benchmarks to test the accuracy of your model(s). It is important to underline that, while very useful to check the compliance of AI applications, benchmarks cannot be simply used in a risk management model. To do so, statistical metrics such as RGA are necessary.

However, ensuring the use of a benchmark or multiple benchmarks to test the compliance for various Responsible AI principles is best practice and highly recommended.

How Will You Communicate This As Needed?

Communicating the performance and accuracy of an ML model could be carried out in various ways, ranging from a research paper uploaded to arXiv or a similar platform and presenting at a leading ML conference. Also, ensuring performance is communicated in a clear, concise, and simple manner on the website/application where applicable. This is important as it addresses "transparency" which is an important Responsible AI principle and could help improve the trust index of consumers and users.

Scoring Rubric

A scoring rubric assists with performance assessment against a standard set of criteria. It acts as a tool to guide ML practitioners on the performance of their ML model(s) in relation to a particular ML metric. In this case, we'll apply a scoring to the various Responsible AI principles outlined in this book. Let's look at a very simple rubric for accuracy.

	Excellent	Good	Fair	Borderline	Poor
Are the outputs of your model accurate?	75%–90%	70%–75%	60%–70%	40%–60%	0%–40%

Mitigation

Now that you have a thorough understanding of the importance of accuracy as a Responsible AI principle and how it could affect the performance of your ML model(s), and ultimately the success of your business, it's important to understand various mitigation and improvement measures to improve the performance of your ML model(s).

The accuracy of your model could potentially make or break your business, for example, if you have an ML model that identifies fraudulent claims and the accuracy of your model is poor, it means that you could be falsely accusing innocent people of fraudulent claims. This could not only damage the reputation of your business and potential/existing customers, it could also lead to reputational damage, lawsuits, and consequently regulatory fines.

Depending on if you have a poor accuracy score from the above rubric, the next step is to optimize performance by carrying out the following:

1. Increase the number of data in your training set. The more examples, the more accurate the model's performance will be. The only caveat to note is increased data equals increased cost, unless using synthetic data.

2. Increase the number of variables which improves feature processing. Feature processing refers to the selection, manipulation, and transformation of raw data into features used in supervised learning.

3. Tune your model parameters while considering alternate values for the training parameters used in your learning algorithm.

There are a few additional things to consider such as the model fit. Is it underfitting or overfitting? Your model is "underfitting" when it performs poorly on the training data, and it's "overfitting" when on the flip side, it performs well on the training data, but poorly on the evaluation data. When this happens, it's usually due to memorization of the data it's been trained on, and the model is unable to generalize unseen examples.

Underfitting can be resolved by increasing the model flexibility and reduction of the amount of regularization used. Regularization refers to a set of techniques used to reduce model complexity and increase generalization which improves model accuracy. To solve overfitting, reduce the model flexibility through less feature combinations, n-grams size, or increase the amount of regularization used.[25]

To summarize, in this chapter we looked at "accuracy," a few key questions from NIST's risk management framework to consider when testing for accuracy, RGA which is a unified measurement for accuracy using the "Rank Graduation" measure, ways to evaluate the performance of your ML algorithms and models, a couple of benchmarks, and mitigation recommendations. In the next chapter, we'll look at robustness and how to ensure you have secure and robust ML models.

CHAPTER 3

Robustness

In this chapter, we'll look at "robustness," another important Responsible AI principle and one of the principles in the SAFE-HAI framework. We will look at a brief definition of robustness and how it relates to the reliability of ML models, with specific reference to NIST's risk management framework. We will also discuss the metrics for measuring robustness, with an introduction to the Rank Graduation "Robustness" score (RGR). We'll conclude this chapter with a brief analysis of an adversarial benchmark for robustness, recommendations for mitigation, and a scoring rubric that could be adopted to analyze the robustness scores of your ML models/ applications.

Robustness is another fundamental and important principle of Responsible AI and AI performance in general. Robustness refers to the sturdiness of an ML model to withstand uncertainties (perturbations or adversarial conditions) such as cyber attacks, and the ability to perform accurately in different contexts. Given the current concerns and unending systemic risks identified in AI systems, particularly generative AI, building robust ML models is not only advised but critical.

According to NIST's RMF, an AI system has to be trustworthy. To achieve trustworthiness, one of the important characteristics is to have "valid and reliable" AI systems. These are reliable systems that have the ability to perform as intended without potential failure over a given period of time, under specific conditions. Another important component of trustworthy AI is the reliability of an AI system to be correct and work as

T. Duke and P. Giudici, *Responsible AI in Practice*,
https://doi.org/10.1007/979-8-8688-1166-1_3

intended during its entire life cycle/lifetime. As such, accuracy (covered in the previous chapter) and robustness are basic components to ensure the validity and reliability of AI systems.

Let's take a look at the NIST's definition of robustness - the ISO/IEC TS 5723:2022 defines robustness or generalizability as the "ability of a system to maintain its level of performance under a variety of circumstances."

In NIST's risk management framework, robustness, along with accuracy, is a key characteristic an AI application/system should have to be considered trustworthy and, specifically, reliable. Reliability is defined as the "ability of an item to perform as required, without failure, for a given time interval, under given conditions." While a valid AI application should be accurate, a reliable AI application must be robust.

Robustness is not only the key principle behind reliability but is also pivotal for other two trustworthy characteristics in the NIST risk management framework, namely, that an application should be secure and resilient, as well as safe. The two aspects are distinct, but closely related. Safety aims to avoid harms that affect the health of people, property, and the environment. Security, on the other hand, aims to avoid unauthorized access of data such as adversarial attacks and data poisoning. Security is directly related to the notion of robustness and specifies its consequences in terms of reliability of an AI application, that is, the expectation that it works as expected, whereas safety can be seen as an indirect consequence of lack of robustness, which leads to unexpected harms on people and the environment. In this chapter, we will focus on the measurement of robustness and the "technical" notion that underpins reliability, security, resilience, and safety.

The EU AI Act also refers to robustness and the requirement to ensure that high-risk AI systems have an appropriate level of robustness and cybersecurity in order to perform consistently throughout the life cycle of an AI system. It's required that AI systems are resilient against errors, faults, and inconsistencies that could occur in the model/application, and appropriate technical measures and solutions are included to backup or "fail-safe" plans especially against malicious attacks against the training dataset, for example, data poisoning, or pretrained models, for example, model poisoning, or adversarial attacks aimed at manipulating the results/output of a model which could include data leakages and confidentiality attacks.

NIST's RMF addresses the following question which also aligns with the EU AI Act:

– How will we protect the AI system against cyber attacks, adversarial attacks, data poisoning, model leakage, evasion, inversion, etc., and ensure ongoing performance? How will we ensure the system is robust to optimizers that aim to induce specific system responses?

The following sections in this chapter will address this critical question.

Measuring Robustness

The measurement of robustness draws on the variation of the output of a machine learning model caused by data perturbations, generated intentionally (as in cyber attacks) or unintentionally (as in extreme rare events). A model is robust when the variation of its output under a data perturbation is limited, and it's not robust otherwise. A more precise definition of robustness can be stated for classification problems: a classifier is robust if it correctly classifies a large proportion of cases, under all possible variations of the input variables (within a certain distance, see, e.g., Croce et al., 2023).[26]

Typically, simpler models with fewer explanatory variables are more robust than more complex models, with more feature variables, as the sources of anomalous variation are less in the former case. This aspect is also known as the bias-variance trade-off (see, e.g., Hastie et al., 2023),[27] where simple models have more bias than complex models, but lower variance. In the measurement of robustness, we will thus consider two apparently distinct but related aspects: a direct measurement of robustness (ex post), in terms of output variability under data perturbations, and an indirect measurement of robustness (ex ante, by design), in terms of model comparison and regularization, which leads to the selection of a model with an appropriate bias-variance trade-off.

The development of this chapter is based on two main benchmark metrics: the RGR metrics, introduced by Giudici and Raffinetti (2024),[28] which measures the "distance" between the predicted output and the predicted output under perturbations, for models based on structured data, and the RobustBench metrics, developed by Croce et al. (2023), which measures a similar distance, for large language models based on unstructured data.

Model Robustness

In this section, we consider how to measure the robustness of a model under perturbations, regardless of their source, that is, whether they are intentional or not. These measurements are also applicable to cybersecurity, security, and safety (indirectly).

The simplest way to measure robustness is to consider the variability of the predictions. For a continuous response, we can resample the train-test partition and obtain repeated predictions, one for each sample. We can then calculate the mean of the predictions, as well as their variance, which measures how robust the predictions are by changing the underlying samples. We can also calculate percentiles of the predictions, which can be used to support decisions.

We recall that the p-th percentile of a statistical distribution X, indicated with xp, can be obtained from the following equation, in which "Freq" indicates the frequency (whose limit is the probability):

$$\text{Freq}\ (X<-xp) = p$$

Or equivalently

$$\text{Freq}(X > xp) = 1-p$$

For instance, the 5th percentile, x0.05, is obtained setting

$$\text{Freq}\ (X<-x0.05) = 0.05$$

Or equivalently

$$\text{Freq}(X > x0.05) = 0.95$$

The obtained 5th percentile x0.05 can then be used as a threshold. For example, if we repeat machine learning on different samples, to predict the value of a specific observation, the prediction will be significantly higher than zero if the 5th percentile is greater than zero, which means that at least 95% of the predicted values are above zero.

For a binary or ordinal response, we can use the entropy or the Gini index, instead of the variance, which are both measures of variability for categorical variables.

We can exemplify the point with an example, taken from a paper written by Babaei and Giudici (2024).[29] The authors consider a situation in which a platform (Lending Club) assesses, based on a machine learning model, whether a set of individuals should be given credit or not – considered a binary classification problem. They compare a "classic" machine learning model, based on a logistic credit scoring model, with the predictions obtained using the Generative Pretrained Transformer model (GPT) 3.5, both without training and with 80 examples (against 30,000 employed by the logistic regression model).

Table 3-1 shows the results of the comparison, comparing the AUROC obtained from five different training samples, in which the classifications to be predicted are the same (170 individuals).

Table 3-1. *Comparison of performance of three models using different training samples*

	Min AUC	Max AUC	Mean AUC
Classic model	0.70	0.79	0.75
GPT	0.59	0.65	0.61
Informed GPT	0.63	0.70	0.67

Table 3-1 shows a GPT model without training examples, outperforming a random model, with a mean AUROC of 0.61. When informed with a minimum set of examples (80), it already reaches a mean of 0.67. On the other hand, a classic model has an accuracy of 0.75, but obtained with 30,000 examples! It's important to note that the range of all three models (max-min) is quite large, especially when examples are being considered (e.g., for the classic model and for the informed GPT).

This indicates that machine learning models, and particularly generative AI models, which are based on large training samples, may have a high variability (a low robustness) and may therefore be extremely sensitive to extreme and/or cyber manipulated data.

However, the method described above is based on the available data and may not be able to provide a measure of robustness of the predictions under perturbations in the data, especially when extreme perturbations lead to data that have not been observed in the available sample. To this aim, we propose to extend the Rank Graduation Accuracy (RGA) measurement to a Rank Graduation Robustness (RGR) measurement, comparing the predictions obtained with the model with those that would be obtained under data perturbations. While the RGA is based on the difference between the observed cumulative response values calculated respectively using the observed and the predicted ranks, the RGR is based on the difference between the observed cumulative predicted values calculated, respectively, using the available data and the perturbed data.

More precisely, the summary of robustness metrics for a model can be obtained by considering the area between the dual Lorenz curve and the concordance curve, derived using the ranks of the predictions obtained with perturbed data, and dividing it by its maximum possible value – the area between the dual Lorenz curve and the concordance curve. This is considered the Rank Robustness (RGR) measure.

Mathematically, the Rank Graduation Robustness measure (RGR) is defined by

$$\mathrm{sum}(Q'i-QPi)/\mathrm{sum}(Q'i-Qi),$$

where Q'i is the dual Lorenz curve, Qi is the concordance curve obtained using the ranks of the predictions, and QPi is the concordance curve obtained using the ranks of the perturbed predictions.

The RGR measure has some important properties, in particular:

1) It is normalized: RGR=0 in the worst case of a nonrobust model; RGA=1 for the best case of a perfectly robust model; otherwise, 0<RGA<1.

2) To assess whether the RGR metric passes a set threshold indicating a minimum level of accuracy, a statistical test for RGR has to be made. When the response Y is binary, the test coincides with DeLong's test. In general, a statistical test based on U-statistics can be derived. In all cases, the higher the RGR, the higher the predictive accuracy, and the smaller the p-value, the more significant is the found RGA with respect to a given lower bound.

Let's exemplify the calculation with a numerical example. Consider the predictions obtained in Table 2-3 of Chapter 2, and suppose a cyber attack that poisons the input data of the input (explanatory) variables occurs, resulting in the perturbed predictions in the third column of Table 3-2.

Table 3-2. *Calculation of the RGR measure for a predictive problem*

Individual	Observed Response	Predicted Response	Qi	Qi'	Predicted Perturbed	Perturbed Ranks	QPi	Q'i-Qi	Q'i-QPi
1	50	70	0.14	0.26	110	4	0.20	0.12	0.06
3	100	80	0.3	0.5	80	1	0.44	0.2	0.06
2	80	100	0.5	0.7	100	3	0.60	0.2	0.10
5	150	120	0.74	0.86	120	5	0.70	0.12	0.16
4	120	130			90	2			

From Table 3-2, it shows that the sum of the differences of Q'i-QPi is equal to 0.48, whereas the sum of Q'i-Qi is equal to 0.64, from which the value of RGR is 0.48/0.64=0.75, indicating a reasonable level of robustness of the predictions against this perturbation. Looking at the data, this is due to the fact that the main changes in the predicted ranks occur for individuals 1 and 4 whose ranks become fourth and second vs. first and fifth according to the predicted model.

The exercise can be repeated many times, possibly considering a set of sample perturbations of the input variables and the mean RGR calculated. An AI application will be deemed robust if the average RGR is significantly higher than a given bound (determined by a set risk appetite). We can also evaluate whether the perturbation significantly deteriorates the prediction obtained with the original data, with a similar resampling test, which is implemented in the Python code referenced in the case study and provided in the appendix of the book.

Let's now consider, as a further example, a classification problem. This allows us to provide an intuitive explanation for the very important case of classifier robustness to cyber attacks, much considered in the literature on adversarial robustness (see, e.g., Croce et al., 2023, for a review). This widely debated stream of research considers both the case of white-box attacks (for which the machine learning model is known) and that of black-box attacks (in which the machine learning model is not known). In both cases, the aim of the attack is to alter the model classification based on the predicted values without perturbations. The literature also considers the case of generic attacks aimed, for example, at corrupting the available input images or texts. This situation can be analyzed as previously described for predictive problems, with continuous response variables.

Similar to accuracy, we suggest using the same metrics as for the continuous case: the RGR measure. The advantage of using a unified measure is evident also in this case – the robustness of the model to different cyber attacks can be measured in the same fashion regardless of the aim of the attack by altering the classifications or more simply corrupting the predictions.

Here's a numerical example for this. Consider the classifications obtained in Table 2-6 in Chapter 2, and suppose a cyber attack that poisons the input data of the input (explanatory) variables occurs, resulting in the altered classifications in the third column below. Specifically, we suppose that the cyber attack alters only the prediction for the individual n.2, with the aim of altering its classification (e.g., bringing it closer to a "1" classification, rather than a "0").

From Table 3-3, we obtain that the sum of Q'i-Qi (maximum possible perturbation) is equal to 1.31, whereas the sum of Q'i-QPi (actual perturbation effect) is equal to 0.89, leading to an RGR equal to 68%, a value which indicates limited robustness, which requires attention.

Table 3-3. *Calculation of the RGR measure for a classification problem*

Individuals	Observed Response	Predicted Response	Qi	Qi'	Predicted Perturbed	Perturbed Ranks	QPi	Q'i-Qi	Q'i-QPi
2	0	0.10	0.04	0.32	0.60	3	**0.19**	**0.28**	**0.13**
5	1	0.46	0.21	0.58	0.46	1	**0.34**	**0.37**	**0.24**
1	0	0.52	0.41	0.79	0.52	2	**0.50**	**0.38**	**0.29**
4	1	0.71	0.68	0.96	0.71	4	**0.73**	**0.28**	**0.23**
3	1	0.85			0.85	5	**1**		

We conclude this section with the Python code required to calculate the RGR, for any given machine learning application.

The function is called

check_robustness

It provides different functions related to the calculation of the Rank Graduation Robustness measure (RGR).

The RGR function can calculate the Rank Graduation Robustness (RGR) of a model, for classification (categorical response) or for prediction (continuous response). It requires a preliminary setup of the type of data perturbation that will be applied to the model. For example, we can consider extracting from the available highest values (e.g., higher than the 95th percentile) of an explanatory variable and replace them with the lowest values (e.g., lower than the 5th percentiles) and vice versa.

Then, the function rgr.single (variable, perturbation.percentage=0.05) perturbs the training values of a (single) selected explanatory variable swapping the values in the 5% lower tail of the corresponding distribution with the values in the 5% upper tail. The default value of 5% can be changed.

Alternatively, the user can use the function rgr.all(perturbation. percentage=0.05) which perturbs in a similar fashion all explanatory variables.

More details on the Python implementation, including a statistical test for RGR, which provides a corresponding p-value are provided in the appendix.

Model Comparison

To improve the robustness and cybersecurity of AI applications (their sustainability), we can also work on the structure (the architecture) of a machine learning model, reducing the number of predictors and/or of their functions involved in the model. It is well known that a complex model may improve the accuracy of a model (its bias) but at the expense of a higher variance of the results, leading to possible overfitting. To solve this problem, which helps with the selection of a model with the appropriate bias-variance trade-off, several approaches have been proposed for model comparison. The most important of them can be classified as dimensionality reduction methods, stepwise selection methods, and regularization methods.

Dimensionality reduction methods aim at summarizing the available input variables in a smaller set of variables, which are functions of the original variables. The most known method is based on the principal components, which replaces k variables with p linear combinations of them, where p is much lower than k. In the principal component method, the p linear combinations are chosen to minimize the loss in variability or, equivalently, the loss in statistical information which derives from reducing the number of variables from k to p.

The obtained transformed variables have the advantage of being uncorrelated with each other so that, for example, the explained variance of the response attributed to a model with all components will be the sum of the explained variances of the single components. A second advantage, already mentioned, is the gain in dimensionality reduction, which means a much lower cost in computations and a consequent lower consumption of energy. A disadvantage of this methodology is that the components are combinations of many variables, and it is often not clear how to interpret them especially when they do not have a common physical meaning. It is also important to standardize or normalize the original variables before performing a principal component analysis; otherwise, the variables expressed in a lower scale (higher magnitude) will dominate the others in terms of variability and, consequently, will have a much higher weight in the linear combinations.

When the available variables are not numerical, it is not possible to run principal component analysis. However, a similar technique, named correspondence analysis, can be run on the frequencies of the available categories, and linear combinations can be obtained minimizing the reduction in the appropriate distance metric (such as the chi-square distance).

Model selection and regularization methods maintain all original variables and compare all model combinations that can be obtained, which is quite different from principal component analysis. It is important to note that the number of possible models grows exponentially with the number of variables. For example, with three explanatory variables, the number of possible models is 2^(3*2)/2 = 8; with four variables, 2^(4*3)/2= 64; and with five variables, 2^(5*4)/2 = 1024. In general, with p variables, it is 2^(p*p-1)/2.

The two most important strategies for model selection are stepwise model selection and regularization methods.

Stepwise methods can be either forward or backward.

In a forward model selection procedure, models are compared in increasing complexity. That is, we first compare the model with no explanatory variables vs. the best model with one variable; then, if the best model with one variable wins, we compare it with the best model with two variables. If the best model with two variables loses, we stop and finally choose the best model with one variable. Otherwise, we proceed and compare the best model with two variables against the best model with three variables and so on until we stop. Note that the decision of who wins and also the decision of when to stop are determined by the choice of appropriate accuracy metrics. Typically, researchers and practitioners use AUROC for binary classifications and RMSE for predictive problems, with the related statistical tests, such as DeLong and Diebold and Mariano tests, mentioned in Chapter 2, which opt to reject the simpler model when the p-value is small.

In a backward model selection procedure, models are compared in decreasing complexity. We first compare the most complex model with all variables (the full model) with a model that has all variables but one. If the latter wins, we compare it with a model that has all variables but two. If the latter loses, we stop. If the model without two variables wins, we compare it with a model with all variables but three, and so on, until we stop. As in the forward procedure, the decision of who wins and the decision of when to stop are determined by the choice of accuracy metrics, such as AUROC and RMSE, and of the related tests. Backward model selection is computationally more costly (as it starts the comparison from the most complex model), but it has the advantage of taking into account all variables and their interactions.

In stepwise selection models, the best model is chosen a posteriori, after a collection of models have been fit and compared (how many depend on when the sequence is stopped), while in regularization methods, the choice is done a priori, on the basis of an optimization method. The main idea is to fit a full model, with all variables, but estimate its parameters (coefficients) taking into account model complexity. More precisely, when regularization is not applied, a full model is estimated by choosing the parameters that minimize an appropriate distance between the observed and the estimated values, such as the mean squared error.

Mathematically, the parameters are chosen minimizing

sum(y_i-ypred_i)^2,

which is the distance between the vector containing the true values, y, and the vector containing the predicted values, ypred.

When ridge regularization is applied, the function to be minimized is the same distance plus a penalization term that is an increasing function of the sum of the squares parameter values. Doing so, we reduce the space of solutions to those that have a sum of parameters smaller than a set value. This implies that some coefficients will have parameter estimates very close to zero.

Mathematically, with ridge regularization, the parameters are chosen minimizing

sum(y_i-ypred_i)^2+penalty*sum(beta)^2,

which is the distance between the vector containing the true values, y, and the vector containing the predicted values, ypred, plus a penalty term which multiplies the sum of the squares of the parameters of the model.

When lasso regularization is applied, the penalization term is instead an increasing function of the sum of the absolute values of the parameter values. This implies that some coefficients may actually become zero, thereby reducing ex ante the number of explanatory variables.

Mathematically, with lasso regularization, the parameters are chosen minimizing

sum(y_i-ypred_i)^2+penalty*sum(|beta|),

which is the distance between the vector containing the true values, y, and the vector containing the predicted values, ypred, plus a penalty term which multiplies the sum of the absolute values of the parameters of the model.

Using either stepwise or regularization, the result of model comparison will be a selected model which will be more parsimonious and, therefore, more stable against extreme variations (or cyber attacks) in the data – a model that will not significantly lose predictive accuracy with respect to a full model.

We now go back to the choice of metrics that can be used to compare models in a stepwise procedure. When we can assume an underlying probabilistic structure, the difference in likelihood between two nested models can be calculated, and a statistical test, such as the F-test for a continuous response or the chi-square test for a binary response, can be implemented to assess whether the difference is significant. For example, if models are compared in a backward perspective, removing one variable at a time, starting from the fully parameterized model (with the highest likelihood), the test can provide a stopping rule, based on the p-values of the test. This procedure could lead to a model that, while still accurate, is more parsimonious and, therefore, more sustainable than the full model.

When a probabilistic model cannot be assumed, as in many realistic machine learning models, a stepwise procedure can also be undertaken by ordering variables with an explainability criterion, such as their Shapley values. You could then employ a stopping rule based on the comparison of predictive accuracy, either in terms of Diebold and Mariano's test (when the response variable is continuous) or in terms of DeLong's test (when the response variable is binary).

The stopping rule will add variables as long as predictions are different, that is, until the difference in predictive accuracy between two consecutive models is large and the p-value smaller than a set threshold.

Instead of using separate metrics and tests for the binary and continuous case, we can also use the RGA metrics proposed in the last chapter and the related test, as in the Python code that is provided in the appendix.

We now exemplify the application of RGR for model comparison. Suppose we need to compare three alternative models - a linear regression model, a regression tree model, and a random forest model - on a set of simulated data, to which we apply random perturbations characterized by the 15% of lower and the 15% of upper outliers. While the linear regression model is explainable by design and the regression tree can be explained by visualizing the tree, the random forest is a black-box machine learning model.

We aim to assess whether the degree of robustness associated with each model when they are contaminated with the same anomalous observations produces significantly different results. The obtained findings are displayed in Table 3-4.

***Table 3-4.** Model comparison based on the RGR measure*

	Model Comparison		
	Linear regression	Regression tree	Random forest
RGR	0.7976	0.8137	0.9683

From Table 3-4, note that the random forest model appears more robust than both the linear regression model and the regression tree model, with the RGR measure achieving a value which is close to one. We further investigate this conclusion by checking whether the robustness degree associated with the random forest model is significantly different with respect to that related to the linear regression model and the regression tree model. The RGR-based statistical test provides a p-value smaller than 0.001 when comparing the linear regression model with the random forest model and also when comparing the regression tree model with the random forest model. This implies that the random forest model is actually more robust than the other two competitors. Based on these findings, it seems that, despite their black-box nature, random forest models provide more stable outcomes than white-box models.

With the purpose of further investigating the efficiency of our proposal, we increase the number of predictors from one to four in order to evaluate if increasing the complexity may affect the models' performance in terms of robustness.

To do so, we generate data from a five-dimensional Gaussian distribution with a prespecified variance-covariance matrix and vector of mean values. More precisely, we suppose that the variable which is selected as a target variable, Y, is associated with the remaining variables according to a common correlation coefficient, which can take four possible values: rho=(0.6,0.4,0.3,0.1). Based on these settings, we simulate the data and fit three alternative machine learning models – linear regression, regression trees, and random forests – and we compute the corresponding predicted values.

Similarly to what we previously described, we apply the scenario characterized by a perturbation of 15% of lower outliers and of 15% of upper outliers, this time to all four predictors.

The findings provided by the application of the linear regression model, the regression tree, and the random forest model to the new simulated data are displayed in Table 3-5.

Table 3-5. *Model comparison based on the RGR measure for a higher dimensional problem*

	Model Comparison		
	Linear regression	Regression tree	Random forest
RGR	0.7449	0.9395	0.9782

Table 3-5 confirms the previous findings: linear regression is the least robust model, and the random forest is the most robust. Note also that while the robustness of linear regression and random forest does not change sensibly increasing model complexity, that of the tree models gets much worse. This is likely due to the "local" behavior of tree models, which rapidly increase complexity when even a small number of observations behave differently from the others.

The findings provided by the application of the linear regression model, the regression tree, and the random forest models show that the linear regression model improves its robustness when model complexity increases, whereas the regression tree considerably worsens its performance. Finally, the random forest model presents only a slight worsening, preserving its robustness to the presence of outlying observations. To sum up, the application shows that random forest models are more robust than the other two models and are thus recommended for the considered problem.

More generally, random forests, and similar ensemble models, such as gradient boosting, can provide a good trade-off between accuracy and robustness, which are often at odds, especially for highly complex neural network models: very accurate, but with a limited robustness.

Adversarial Robustness Benchmark

As with any ML model launch or application, the use of benchmarks is always highly recommended and is applicable to robustness. Adversarial robustness refers to the model's ability to resist being fooled by malicious actors. Croce et al. introduced "RobustBench: a standardized adversarial robustness benchmark" (cited above) which evaluates adversarial robustness using AutoAttack, a combination of white- and black-box attacks. The benchmark consists of the following:

- An open source website with a leaderboard on GitHub which contains over 120 evaluations of models, aimed at reflecting the current state of the art (SOTA) in image classification models on a set of well-defined tasks, among others. Users are able to track the progress of the evaluations and the current SOTA based on a standardized evaluation using AutoAttack.
- The library also provides access to over 80 models (model zoo) to facilitate downstream applications, which facilitates the development of improved adversarial attacks as ML engineers can perform evaluations on a large set of over 80 models.
- RobustBench also analyzes the "impact of robustness on the performance on distribution shifts, calibration, out-of-distribution detection, fairness, privacy leakage, smoothness, and transferability," based on the collection of models from the model zoo.

While performing robustness evaluations is highly recommended and crucial to the success of an ML model/application, it's important to note that there are trade-offs with accuracy, where adoption of adversarial robustness could affect performance, leading to severe accuracy penalties. In early 2023, Bai et al. introduced a way to alleviate these trade-offs through "adaptive smoothing."[30] Adaptive smoothing mixes the output probabilities of a standard classifier and a robust classifier while optimizing for clean accuracy. The method includes adapting an adversarial input detector into a mixing network which adjusts the mixture of the two base models - the robust base classifier and the standard classifier. This flexible mixture-of-experts (MoE) framework has proven to reduce the accuracy penalty of achieving robustness, further leading to clean accuracy, robustness, and adversary detection. Adaptive smoothing also uses AutoAttack to evaluate robustness. As discussed for accuracy in the previous chapter, benchmarks are important to assess compliance, but they cannot be simply adoped in a risk management model; for that purpose, a statistical metric, such as RGR, is necessary.

Scoring Rubric

Once the outlined measurement metrics for RGR have been completed, it's important to score the robustness of your model(s). Here's a simple scoring rubric for robustness.

Table 3-6. *A scoring rubric for robustness*

	Excellent	Good	Fair	Borderline	Poor
What is the robustness score of your model(s)?	75%–90%	70%–75%	60%–70%	40%–60%	0%–40%

Mitigation

Now that you have a thorough understanding of how to measure the robustness of your model(s), including a component analysis, and you're able to run several benchmarks to test the robustness of your model, it's important you work on improving the robustness of your model if it comes up with a low score.

Depending on the issue, there are various ways to improve the robustness of your ML model. A few recommendations have been mentioned such as a model selection and evaluation. Here are some additional recommendations:

1. **Have a thorough understanding of the data**: Conduct exploratory data analysis, identify potential outliers, understand the relationships between the various variables in the dataset(s), and gain some overall understanding of the structure of the data use.

2. **Conduct data preprocessing**: This is an important step in the ML pipeline and includes tasks such as normalizing numerical variables or dealing with class imbalance in the target variable. Data preprocessing makes the data compatible with your selected ML algorithm and helps to improve the algorithm's ability to uncover meaningful patterns.

3. **Adopt feature engineering**: This creates new features from existing ones and has a significant impact on the performance of an ML model, helping to capture important aspects of the data that the model may not have identified, further improving the model's predictive power.

4. **Perform cross-validation**: Cross-validation is an important technique that helps to prevent overfitting, where the model memorizes data and performs poorly on unseen data (also explained in Chapter 2). To avoid overfitting, divide the data into training and validation sets several times, which

helps to ensure the model generalizes well on unseen data. As accuracy and robustness go hand in hand, this technique will also invariably improve the robustness of a model.

5. **Keep up to date with the latest research developments**: Keeping up to date with the latest research in the ever-evolving field of ML will help you learn the newest research methods to adopt, ensuring more robust models – including the robustness of generative AI models, which have much more amounts of data and have been "jailbreaked" quite often.[31]

In this chapter, we covered the meaning of robustness and its importance and how to measure the robustness of ML models using the RGR value with comparisons between a random forest model and a linear regression model. We also discussed benchmarks for robustness and ways to improve the robustness of your models if they have a poor robustness score. In the following chapter, we'll look at "explainability" and the importance of developing explainable and transparent ML models.

CHAPTER 4

Explainability

Let's look at the next Responsible AI principle and SAFE-HAI component we propose in this book – explainability.

While the principles of accuracy and robustness have been present for a relatively long time span, the notion of explainability is more recent. Its importance has grown in parallel with the development of "black-box" learning models that may be very accurate, although very difficult, or impossible, to interpret, especially in terms of their driving mechanisms. While a precise mathematical definition of explainability is not available due to its somehow subjective interpretation, the extant literature relies on regulatory definitions, such as that reported in Bracke et al. (2019),[32] where explainability requires an explicit understanding of the factors which determine a given machine learning output.

According to NIST's RMF, an AI system has to be explainable and interpretable. Explainable and interpretable artificial intelligence systems offer information that can help users understand the purposes and impact of AI systems, while interpretability refers to the meaning of AI output in the context of its designed purpose. Explainability refers to the mechanisms which underlay AI systems. The underlying assumption is that a risk of explainability or interpretability stems from a lack of ability to make sense of a system's output.

To achieve trustworthiness, AI systems must be explainable and interpretable. As the two terms are closely related, in this chapter we will use the term "explainability" to mean both.

T. Duke and P. Giudici, *Responsible AI in Practice*,
https://doi.org/10.1007/979-8-8688-1166-1_4

The EU AI Act also refers to explainability, which also embodies transparency. The act states that "High risk AI systems should be designed and developed in a way to ensure their operation is sufficiently transparent to enable deployers to interpret the system's output and use it appropriately." The notion of transparency adopted by the AI Act is clearly related to the notions of explainability and interpretability as previously defined. Another reference of the AI Act to explainability can be found in the requirement of "human oversight." "AI systems must be designed and developed in such a way that they can be effectively overseen by natural persons during the period in which the system is in use." We will refer back to this in Chapter 8.

In the machine learning research community, the growing importance of the explainability requirement has led to an increase in publications and workshops, including conferences dedicated to explainable artificial intelligence (XAI), among which is the World Conference on Explainable AI.[33] The conference solicits papers on the topic of explainable AI and has so far selected an average of 30% submitted papers: a high selectivity which indicates, along with the large number of submissions, that the conference has become the reference forum for researchers that work on the topic of explainable AI and want to learn new methods and new applications, leading to cross-fertilization and improvement of the research.

Indeed, some machine learning models are explainable by design and don't need further refinements to be explainable. For example, linear and logistic regression have made recent advances thanks to feature selection optimization methods such as Lasso and Ridge. Other models such as tree models, random forests, bagging, and boosting models are much less explainable. However, these models have a built-in mechanism referred to as the feature importance plot that calculates the importance of each variable in terms of the splits (or average splits) it generates on the tree (or forest) model. We also have more complex models such as neural networks and support vector machines that are typically unexplainable. This problem, however, can be solved with some computational efforts, by permuting and/or reshuffling the values of any given explanatory variable.

Further research development has been motivated by the need to compare the explainability of different machine learning models, such as regressions, tree model, and neural networks, in terms of an independent model-agnostic measure run in a post-processing step. The most important of such measures are Shapley values (see, e.g., Shapley, 1953,[34] Lundberg, 2017[35]) and their normalized variant, the Shapley Lorenz values (Giudici and Raffinetti, 2021[36]).

Shapley values were introduced in game theory with the aim of dividing the value of a game equally between the various participants. The goal was to extend the technique to machine learning models. In this context, Shapley values allow for the equitable distribution of a model's prediction among the regressors that constitute it.

Let's take a look at a quick game analogy to understand the Shapley game theory. Assuming there is a game for each observation you plan to predict. For each game, the players are the model predictors, and the total gain is equal to the sum of each predictor's contribution. For any given observation, the (local) effect of each variable k can be calculated as follows:

Shapley_i= Sum_c (ypred(Xc) - ypred(Xc\k)),

where the Sum is taken over all c = 1 ,..., C possible combinations of the available explanatory variables and where ypred(Xc) and ypred(Xc\k), respectively, indicate a prediction obtained using the variables in a specific combination and the predictions obtained using the same variables (minus variable k).

In other words, Shapley value measures the added value of a variable, k, to the prediction of each observation.

Once Shapley values are calculated, the global contribution of each explanatory variable to the predictions is obtained by summing or averaging them over all observations.

Shapley values help developers understand the impact of the various explanatory variables in a model's output. In other words, they help you understand which explanatory variable drives the model's predictions.

Unlike various regularization techniques, such as Lasso and Ridge, which are applied in model building (ex ante), Shapley values are applied once a model is selected (ex post). While Lasso and Ridge are mainly used for model selection, Shapley values are employed for model interpretation.

However, it may be worth examining how much variability of the response is explained by a given set of predictors. If a smaller set of predictors is sufficient, we can choose that, making the model more robust to input variations.

To this aim, Giudici and Raffinetti (2021) introduced Shapley Lorenz values, which replace the payoff for Shapley values with a payoff based on the difference between Gini coefficients (Lorenz Zonoids), normalized by definition.

Shapley values and Shapley Lorenz values are examples of post-processing techniques that can be employed to understand whether a machine learning model is explainable and, if so, what are the variables that mostly explain the variability of the response to be predicted. While appealing, these measures, and similar post-processing techniques, may be computationally complex, especially when many variables are being considered. This is because the large number of possible combinations C becomes rapidly very large, as the number of variables increases.

To improve the assessment of explainability, in the next section we will propose a measure based on the Lorenz curve notion, similar to the accuracy and robustness metrics discussed in the previous chapters. This will allow us to answer the following question:

- How will we ensure that an AI system is explainable? How can we measure when an AI system is not explainable?

The following sections in this chapter will address this critical question by means of the Rank Graduation Explainability measure (RGE).

Measuring Explainability

In this section, we consider how to measure the explainability of a model or, more precisely, to understand which explanatory variables drive the response variable predictions, if any. For further review of explainability and interpretability methods, refer to the paper by Saranya and Subhashini (2023).[37]

For illustrative purposes, let's look at the assessment of credit scores for a credit lending platform which analyzes more than 100,000 European companies. The data is described in Figure 4-1.

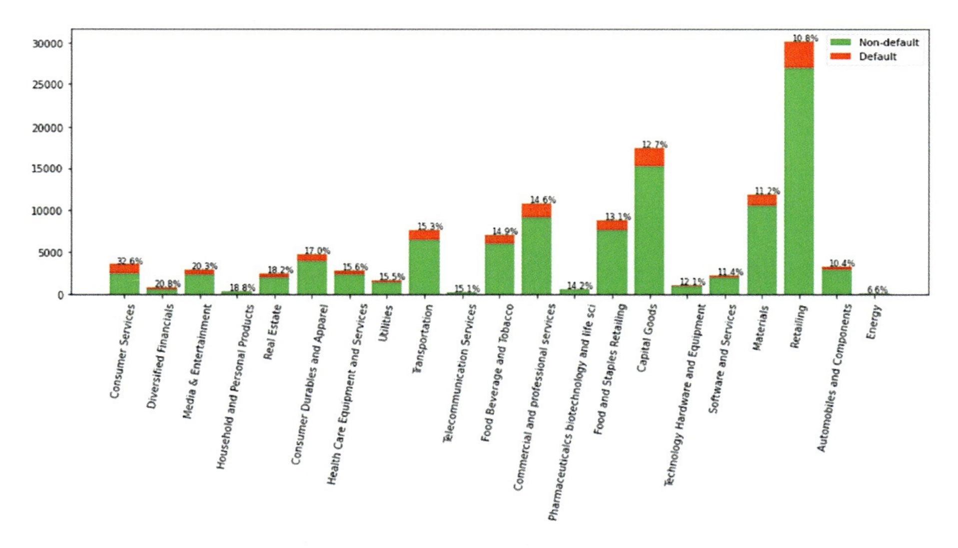

Figure 4-1. *Data distribution of the analyzed companies by sector and default*

Figure 4-2 exemplifies the application of one of the most known methods to obtain variable explanations for the family of ensemble tree models, which contains random forest and gradient boosting, for example, the variable importance plot described below.

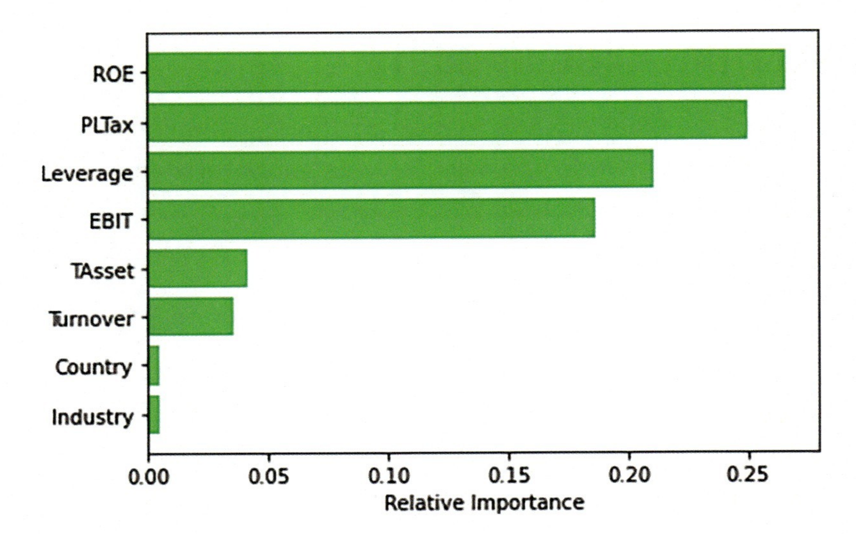

Figure 4-2. *Variable importance plot for the random forest credit scores*

Figure 4-2 is a variable importance plot for the explanatory variables that predict the corporate default in Figure 4-1. They all describe the contribution of a set of financial and demographic characteristics which determine the credit score of a company – an estimate of its probability of default. For each variable, the plot represents its importance, that is, the average reduction in variability obtained using that variable in a random forest of tree models, each time sampling different training/test sets.

Figure 4-2 shows reports for each of the eight variables which describe different characteristics of a company, their importance, as measured by the reduction in heterogeneity of the predictive classifications achieved by the model. The heterogeneity is minimum when all predictions in the same class have the same response values and maximum when they are all different. Some variables, such as ROE, are very important as they can contribute substantially in the reduction of the heterogeneity, when employed to segment the observations into distinct group classes.

The next figure exemplifies the application of Shapley values which, similar to importance plots, are obtained in a preprocessing step, but are agnostic as they can be calculated for all models, not only for ensemble tree models. In Figure 4-3, we present the result of the application of Shapley values to a logistic regression model.

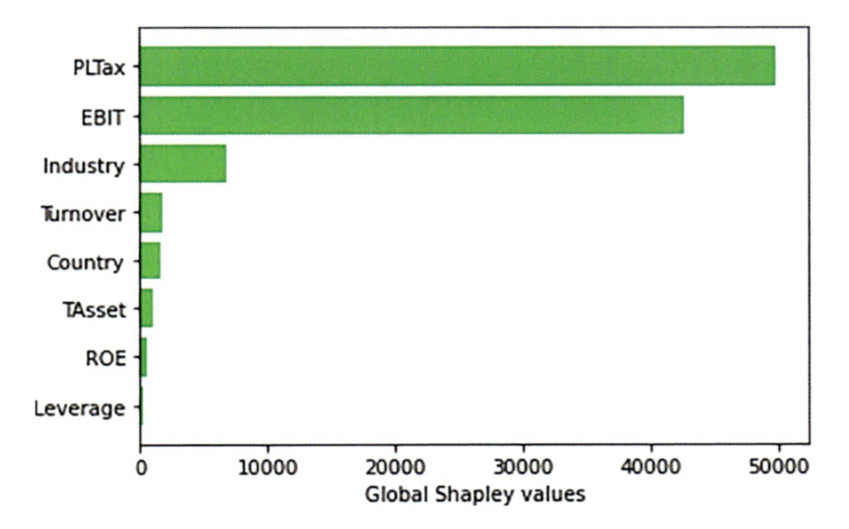

Figure 4-3. *Shapley values for the logistic regression credit scores*

Figure 4-3 represents the global Shapley values, the sum of all Shapley values for the explanatory variables that potentially predict a binary response, the credit default of a company, using a logistic regression model.

Figure 4-3 shows that the variable “PLTax” - which describes the level of profit or losses of a company before taxes - is the variable that most contributes to reducing the heterogeneity of the predicted classes, in terms of defaults/not defaults contained in them. PLTax is followed by EBIT, a variable similar to PLTax, which describes “operating” profits before taxes, but also before financial interests and depreciation of holdings.

Referring back to the previous chapter on accuracy, we saw that the AUROC of the random forest model is higher than that of the logistic regression, which is more accurate. However, the explanations of the logistic regression are "more concentrated," that is, fewer variables matter. Also, recall that the logistic regression model is "explainable by design," so we do not need to calculate Shapley values in a post-processing step as logistic regression already produces estimates for the coefficients of the explanatory variables. This shows a trade-off between accuracy and explainability: in terms of the first principle, random forest wins; but, in terms of the second principle, logistic regression wins.

We can consider the third principle: robustness or sustainability. For instance, we can try to simplify the full model with all variables, for both models, to check if we can obtain a simpler and more robust model. We did model comparison by means of a model comparison procedure based on DeLong's test. The results, reported in Babaei et al. (2023),[38] indicate that, while the random forest model can be simplified removing the variables "industry" and "country," the logistic regression model cannot. The random forest model is thus superior also in terms of robustness. Before drawing to a final conclusion, we will compare the two types of models in terms of fairness in the next chapter.

Model Explainability

As previously discussed, the methods described above may be too computationally complex, especially when many explanatory variables are present in the data. Furthermore, it is necessary to embed them in a statistical testing procedure, in order to understand when a certain variable significantly explains the response (or, in other words, is as a significant explanation). To this aim, let's extend the Rank Graduation Accuracy (RGA) measurement to a Rank Graduation Explainability (RGE) measurement, comparing the predictions obtained with a model based on all explanatory variables with those that would be obtained with a model that excludes the variable of which we would like to measure its explainability. While the RGA is based on the difference between the observed cumulative response values calculated respectively using the observed and the predicted ranks, the RGE is based on the difference between the observed cumulative predicted values calculated, respectively, using all the variables and all without the considered one.

More precisely, a summary explainability metrics for each variable in a model can be obtained by considering the area between the dual Lorenz curve and the concordance curve, derived from the ranks of the predictions obtained from the simpler model (with one less variable), and dividing it by its maximum possible value - the area between the dual Lorenz curve and the Lorenz curve, obtained using the predictions obtained from the full model (with all variables). This is the Rank Graduation Explainability (RGE) measure.

Mathematically, the Rank Graduation Explainability measure (RGE) is defined by

$$\mathrm{sum}(Q'i-QPi)/\mathrm{sum}(Q'i-QFi),$$

where Q'i is the dual Lorenz curve, QFi is the Lorenz curve obtained using the ranks of the predictions obtained with a full model, and QPi is the concordance curve obtained using the ranks of the predictions obtained excluding the variable under measurement.

The RGE measure has some important properties, in particular:

1) It is normalized: RGE=1 in the case of an irrelevant variable, which provides no explanation of the variability of the response; RGA=0 for the case of a variable that explains all the variability of the response; otherwise, $0<RGE<1$.

2) To assess whether the sum of the RGE metrics, for the variables in the model, passes a set threshold indicating a minimum level of explainability, a statistical test for RGE has to be made. A statistical test based on U-statistics can be derived taking the complement to one of the sum of the RGE values. Thus, the lower the sum of the RGE, the higher the explainability, and the smaller the p-value, the more significant is the found RGE with respect to a given lower bound.

Let's exemplify the calculation with a numerical example. Consider the predictions obtained in Table 2-3 of Chapter 2, and suppose one of the variables is removed from the model, resulting in the predictions in the sixth column in Table 4-1.

Table 4-1. *Calculation of the RGE measure for a predictive problem*

Individual	Observed Response	Predicted Response	QFi	Qi'	QPi	Ranks from QP	QPi	Q'i-QPi	Q'i-QFi
1	50	70	0.14	0.26	112	4	0.20	0.12	0.06
3	100	80	0.3	0.5	90	1	0.44	0.2	0.06
2	80	100	0.5	0.7	100	3	0.60	0.2	0.10
5	150	120	0.74	0.86	115	5	0.70	0.12	0.16
4	120	130			95	2			

From Table 4-1, it shows that the sum of the differences of Q'i-QPi is equal to 0.48, whereas the sum of Q'i-Qi is equal to 0.64, from which the value of RGE is 0.48/0.64=0.75, indicating a moderate level of explanation of the considered variable.

This exercise can be repeated for all the variables in the model, and the sum of the RGE can be calculated. An AI application will be deemed explainable if the complement to one of the sum of the RGE is significantly higher than a given bound (determined subjectively). The test can be carried out by means of the Python code referenced in the final chapter of the book and the appendix.

We remark that a similar exercise can be carried out for a classification, rather than a regression problem.

We conclude this section with the Python code required to calculate the RGE, for any given machine learning application.

The function is called

check_explainability

It provides different functions related to the calculation of the Rank Graduation Explainability measure (RGE).

The RGE function can calculate the Rank Graduation Explainability (RGE) of each variable in the model, for classification (categorical response) or for prediction (continuous response). It can then conduct a statistical test to assess whether the overall explanation is significantly higher than a set threshold or to that of another model.

More details on the Python implementation, including a statistical test for RGE, which provides a corresponding p-value, are provided in the appendix.

Scoring Rubric

Once the outlined measurement metrics for RGE have been completed, it's important to score the explainability measure of your model(s). Here's a simple scoring rubric for explainability.

Table 4-2. A scoring rubric for explainability

	Excellent	Good	Fair	Borderline	Poor
What is the explainability score of your model(s)?	75%–90%	70%–75%	60%–70%	40%–60%	0%–40%

Mitigation

We'll use RGE values to provide mitigation measures for explainability. Unlike RGA or RGR, RGE is not a single statistics but rather a collection of statistics, one for each feature in the model. Therefore, before making mitigation actions on RGE, it is necessary to aggregate the corresponding values. This can be done by examining the distribution of RGE values among the different features. For example, we can make a plot in which the feature variables are ordered in terms of their RGE values, from the highest to the lowest. Then check how many features are needed to reach a total RGE of 50%. If the 50% is reached with a small number of features, the model is explainable. If 50% is not reached after adding the RGE of all features, or if it is reached, but with a very large number of features, the model is not explainable.

The threshold of 50% is purely indicative; a user can set a higher or a lower threshold, depending on the desired level of explainability. In terms of the number of explainable features, the definition of a "small number" again is subjective and context specific. If the context is that of continuous monitoring, "small" may mean three or four features; if the context is more descriptive, "small" may mean ten or twelve features.

If a machine learning model is evaluated as not being explainable, there are a number of actions which can be taken:

1. **Make changes to the model**: The machine learning model can be improved by means of model selection, which means trying to simplify the model by removing the features which are correlated with the others, without a significant reduction in the overall accuracy.

2. **Make changes to the datasets**: Data quality can be improved by removing outlier data, or data that are not representative. Additionally, you can adjust the training data to better represent certain features (by means of oversampling or undersampling). This may lead to improved balance between the features.

3. **Increasing the amount of data**: Explanations can also be improved by increasing the available data in the dataset(s) and augmenting the available data with data from similar problems (meta analysis) or generating synthetic data.

It's important to note that the advantage of our proposed metrics, RGE, with respect to standard measures such as Shapley values, is that it is normalized and interpretable as the percentage of model accuracy due to a feature. This allows us to evaluate the results of mitigation actions directly, in terms of the added accuracy/explainability brought to the model by the proposed changes. This is currently not possible with Shapley values as the differential measurement is not normalized and not connected to predictive accuracy.

In this chapter, we've looked at explainability, a core SAFE-HAI principle, and discussed its recency in relation to accuracy and robustness. We also honed in on its emphasis in NIST's RMF and its importance.

We reviewed Shapley values and the game theory with respect to explainability. To measure explainability values in an ML model, we introduced RGE, the Rank Graduation Explainability measure, an important metric which not only helps to measure explainability scores but can be applied in mitigating and fixing poor explainability measures in ML models. In the next chapter, we'll look at "fairness" and "human rights" as the fifth Responsible AI principles in the SAFE-HAI framework.

PART III

Ethical Risks (External)

CHAPTER 5

Fairness and Human Rights

In the previous chapter, we looked at explainability and its importance as a key Responsible AI principle/attribute. We also discussed ways to measure explainability using the RGE metric and reviewed a few adversarial benchmarks for robustness, including ways to improve the robustness of your models, if they score poorly on the robustness scoring rubric. The next Responsible AI principle we'll discuss is "fairness" and how it relates to human rights. These principles will be covered in this chapter.

The motivation to develop Responsible AI technologies aims at mitigating the risks that AI systems could cause harms on individuals, organizations, and the environment. Such harms may be physical or psychological – which are usually associated and addressed under "AI safety." We also have concerns about the violation of human rights, such as the right of not being discriminated against and the right to privacy, among others. In this chapter, we will focus on the risk of discrimination and how to develop fairness.

The concept of fairness in machine learning has attracted a lot of interest in recent years. Despite the vast number of studies on fairness, there are no universal measures to calculate it, and several notions of fairness have been considered in various literature.

T. Duke and P. Giudici, *Responsible AI in Practice*,
https://doi.org/10.1007/979-8-8688-1166-1_5

For example, Teodorescu (2021)[39] follows a group-based approach to fairness. The paper uses statistical fairness measures and relates the fairness of machine learning models with computational complexity employing an imbalanced large credit scoring data. They show that when more than one protected variable is considered in a classification problem, the typical ML models do not satisfy more than one fairness criterion. This shows the difficulty of calculating fairness in decision-making processes using large databases such as credit scoring. The paper also discusses the problem of selecting suitable algorithms that satisfy specific ethical criteria to classify real data. In particular, they focus on the statistical fairness measures which show that typical classification algorithms are not always fair toward different protected attributes. The results found by analyzing a huge credit scoring dataset prove the need for human input in fairness decisions, especially when deciding trade-offs between fairness criteria.

A second paper by Horesh et al. (2020)[40] considers individual measures of fairness. In this case, there is no explicit protected variable but other correlated demographic features that lead to discrimination and bias in the model predictions. The proposed approach in this paper has the following two general aspects: despite most fairness methods which rely on the presence of at least a protected variable, this proposed approach evaluates individual-based fairness using an implicit idea of a potentially discriminatory variable that does not have to be directly measurable. Secondly, it is applicable to a wide range of ML models. In particular, they define a paired consistency score, which measures the similarity of a model's predictions, both for regression and classification, for paired members.

Other papers focus on an experimental or "counterfactual" design setting. For example, Karimi et al. (2022)[41] propose the "FairMatch" algorithm which introduces a new approach to pairing similar and dissimilar individuals using a propensity score (PSM) matching method. They focus on individual fairness and propose a novel metric to evaluate individual fairness that captures the notion of similar treatment in probabilistic classifiers in a better way. Using four real-world datasets, they show the superiority of FairMatch to the existing approaches. They also use the PSM method not only for matching the similar individuals but also to improve the group-based fairness.

It's important to note that the requirement of fairness is related to the requirement of explainability. A recent paper by Grabowicz et al. (2022)[42] combines the two requirements of fairness and explainability to obtain models which are both fair and interpretable. We follow a similar approach in this chapter. By doing so, we also comply with the line of research that not only measures but also suggests how to remove bias, as discussed in Jiang and Nachum (2024).[43]

Measuring Fairness for Organizations

To better understand how fairness can be operationally assessed, we will discuss the measurement of fairness and explainability using the case study on credit scoring for corporate organizations that we introduced in Chapter 4. In the next section, we will consider a real-life case study that relates to consumer credit for individuals.

Figure 5-1 represents the data we are considering from a different perspective with respect to explainability, discussed in Chapter 4.

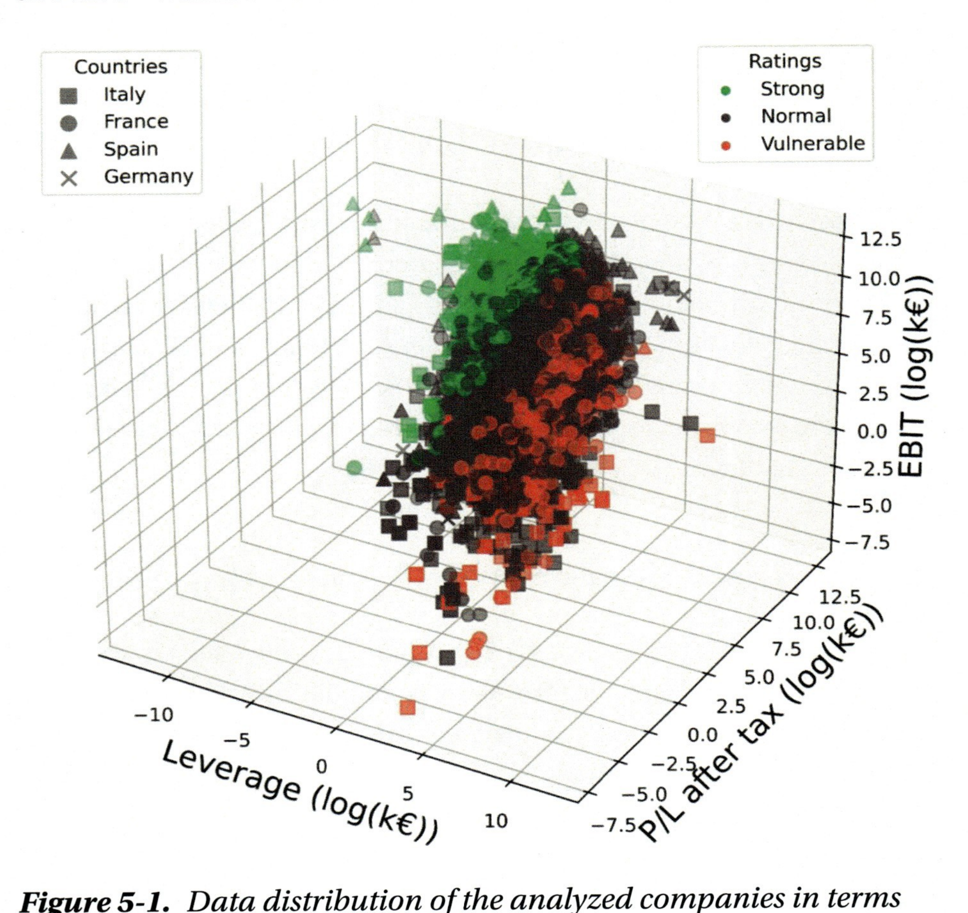

Figure 5-1. *Data distribution of the analyzed companies in terms of three balance sheet characteristics: leverage, profit and losses (P/L) after tax – earnings before interest and taxes (EBIT). Source: Chen et al*[45]

Figure 5-1 represents a set of more than 100,000 small and medium enterprise organizations requesting credit, which are evaluated in terms of their balance sheet results and, particularly, in terms of leverage, profit, and losses – EBIT. The question we address is whether the credit scores are assigned fairly and conditionally on their financial performance – in other words, whether companies with similar financial performances receive similar credit scores.

A natural framework to assess this type of fairness is by means of a Lorenz curve which represents the cumulative distribution of the Shapley values of a variable (such as leverage) for the different considered countries. In Figure 5-2, we depict this curve using Shapley Lorenz values instead of Shapley values.

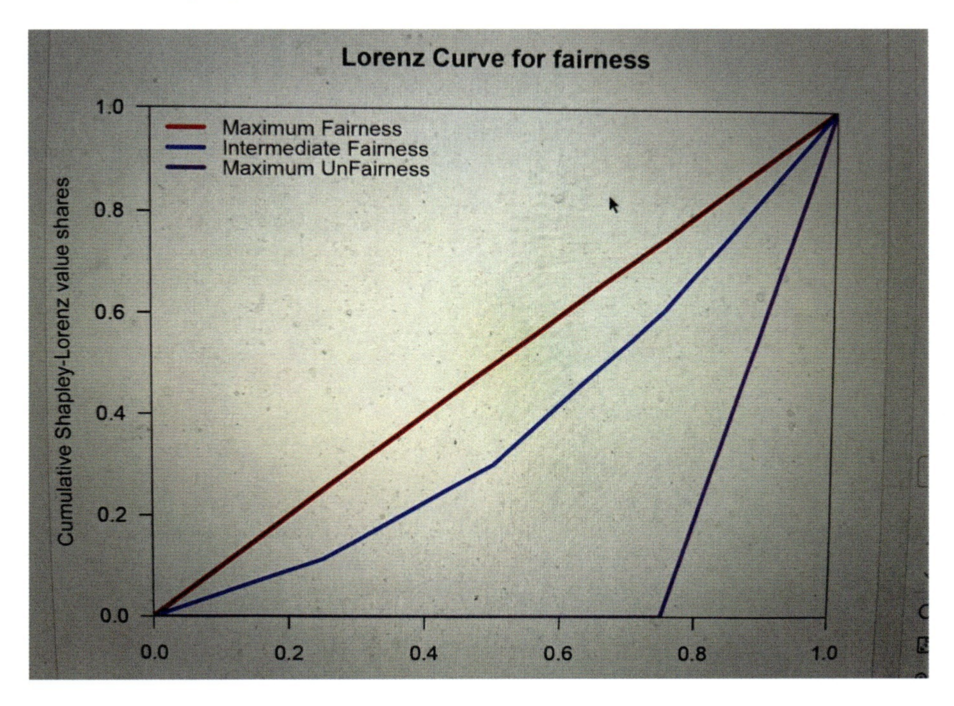

Figure 5-2. *Possible situations for conditional fairness. Source: Chen et al*[45]

Figure 5-2 depicts two stylized situations of minimum and maximum fairness. Among them, the blue curve indicates the real considered case, which indicates a high degree of fairness, being close to the red curve rather than the purple one. Using a notion of conditional fairness where a machine learning model is considered "fair" when a given explanatory variable, such as "leverage," has the same effect on the output. This is different from marginal (unconditional) fairness, which indicates that the output and not the explanation is the same for different population groups. Conditional fairness can be associated with "individual" fairness, whereas marginal fairness can be associated with group-based fairness.

An example of group-based fairness is indicated in Figure 5-3 where we show a machine learning output - the predictive accuracy, for different countries and years.

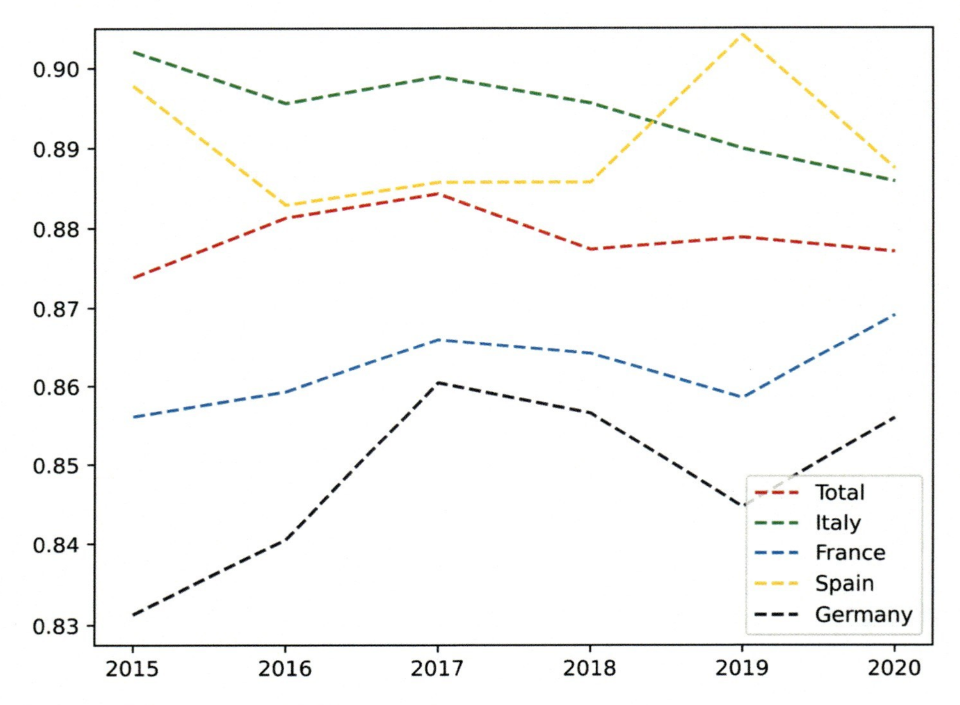

Figure 5-3. *Marginal fairness by countries and years.*
Source: Chen et al[45]

Figure 5-3 indicates a high degree of fairness, as predictive accuracies are similar and vary across the different years.

This analogy poses the question: Which notion of fairness to employ? Marginal fairness is easier to understand. On the other hand, conditional fairness is more useful when there are structural characteristics, corresponding to explanatory variables that make certain biases. For example, in the considered case, German companies are not required to deposit balance sheets, different from the others, and, therefore, the companies in the sample are much fewer. We thus expect a lower accuracy which, however, does not necessarily mean a bias. The Shapley values for leverage in the different countries are reported in Figure 5-2, and they indeed reveal a small bias. The same can be shown for the other variables. Thus, the model is conditionally fair. In the next section, we will see how to assess fairness precisely, using the Rank Graduation Fairness (RGF) metric.

Measuring Fairness for Individuals

In measuring fairness for individuals vs. organizations, which was discussed in the previous section, let's consider consumer lending, a high-risk application. Consumer lending is considered high risk as it could imply unfair allocation of credit to certain individuals.

To discuss Responsible AI methods in this context, we consider a publicly available database, made up of 157,269 loan applications made in 2017 in New York, taken from the Home Mortgage Disclosure Act (HMDA) data repository. The dependent target variable, "declined loan," takes the value of "1" if a loan application initially satisfies the approval requirements of Government-Sponsored Enterprises or of Federal Housing Administrations (GSEs/FHA); it takes the value 0 if the lender approves the loan. Since the Global Financial Crisis in 2008, many lenders have enforced stringent approval requirements besides those of GSEs and FHA. This means that despite satisfying the requirements of GSEs/FHA, an applicant's loan request may still be rejected by the lenders. In detailing GSEs/FHA's initial acceptance of the borrower's application and its subsequent rejection by the bank, the HMDA dataset, which includes information on the applicant's race, eminently qualifies for this example.

Let's review a key independent variable of interest (the protected variable): the information on the applicant's race. It takes the value 1 if the applicant is African American and 0 if it is White American. Further, independent variables are used as controls, including gender, whose variable takes the value 1 if the applicant is male and 0 if female; the applicant's gross annual income; the amount of loan applied for; the purpose of the loan (e.g., 1 for refinancing the mortgage and 0 for purchasing a home); lien status, which takes the value 1 if the loan application is secured by "first lien" and 0 for a subordinate lien; and type of loan, whose variable takes the value 1 if the loan was insured by the FHA and if insured by a GSE.

A preliminary data analysis shows there is a high imbalance in the sample with only roughly 8% African Americans represented. On the other hand, the applicant's race is highly correlated with many other independent variables, and, in particular, it is highly and positively correlated with the "loan amount." The combined effect of such dependencies suggests a possible paradox: credit decisions may be unfair conditionally on some explanatory variables (and, particularly, on loan amount), but fair when not considering them.

In other words, fairness in loan acceptance should be evaluated not only by looking at whether the probability of accepting a loan depends on race (direct bias) but also by looking at whether the same probability changes conditionally on a control variable that is highly correlated to race (indirect bias), such as the "loan amount."

This type of statistical paradox is known as "Simpson's paradox." It is a phenomenon in which a statistical dependence appears in different groups of data but disappears or reverses when these groups are combined. Typically, the paradox arises when there are confounding variables that are not taken into account during the analysis. The observed association or trend in the aggregated data can be misleading because the true relationship is hidden, and it is revealed when the data is disaggregated by relevant factors.

For the data at hand, Agarwal et al. (2023)[44] have shown, using the Shapley Lorenz values of Giudici and Raffinetti (2021), that loan acceptance decisions are more fair when a random forest machine learning model rather than a traditional logistic regression model is employed. Their analysis, however, is marginal and not conditional on the available control variables.

To assess whether loan acceptance decisions are fair, we can compare the Shapley Lorenz values of the explanatory variables, focusing on the applicant's race. This is a conditional approach, similar to that followed in Chen et al.[45]

The authors compare a logistic regression model with a random forest model on 70% of the data and apply the global Shapley method to examine whether there is evidence of racial discrimination dividing the dataset with respect to loan amount: one if higher than the mean amount requested and zero if lower.

The authors found that, in line with the expectations, the random forest model is more accurate than logistic regression. In terms of fairness, when applying a random forest for the high loan amount sample, the applicant's race is the most important explanatory variable. It is ranked first out of our six explanatory variables, and it explains 91% of the declined loans. In contrast, the applicant's race is only the fifth most important variable (loan type is the most important), and it explains only 1% of the variability of declined loans. When the two samples are combined, as in Agarwal et al. (2023), the applicant's race explains only 3% of the response.

This shows clear evidence of a Simpson's paradox: the random forest model appears (marginally) as an ethically accountable model, differently from the logistic regression model. However, when we consider separately the loan amount requested, the model becomes unfair, in particular when a high amount is requested.

In the next section, we present the RGF metric, which is conditional and, therefore, avoids these types of paradoxes, and at the same time, it is easier to compute.

Model Fairness

The methods previously described may be too computationally complex, especially when many explanatory variables are present in the data. To this aim, we will extend the Rank Graduation Accuracy (RGA) measurement to a Rank Graduation Fairness (RGF) measurement, comparing the predictions obtained with a model based on all explanatory variables with those that would be obtained with a model that excludes the protected variable. While the RGA is based on the difference between the observed cumulative response values calculated, respectively, using the observed and the predicted ranks, the RGF is based on the difference between the observed cumulative predicted values calculated, respectively, using all the variables - but the considered protected variable.

More precisely, the fairness metrics for a model can be obtained by considering the area between the dual Lorenz curve and the concordance curve, derived from the ranks of the predictions obtained without the protected variable, and dividing it by its maximum possible value - the area between the dual Lorenz curve and the Lorenz curve. The Lorenz curve is derived using the ranks of the predictions obtained from all variables in the model. This is considered the Rank Graduation Fairness (RGF) measure.

Mathematically, the Rank Graduation Fairness measure (RGF) is defined by

sum(Q'i-QPi)/sum(Q'i-QFi),

where Q'i is the dual Lorenz curve, QFi is the Lorenz curve derived using the ranks of the predictions obtained from the full model, and QPi is the concordance curve obtained using the ranks of the predictions derived by excluding the protected variable under measurement.

The RGF measure has some important properties, in particular:

1) It is normalized: RGF=0 in the case of unfairness – the protected variable reverses the ranks; RGF=1 for the case of fairness – the protected variable does not change the ranks; otherwise, 0<RGF<1.

2) To assess whether the RGF metric passes a set threshold indicating a minimum level of accuracy, a statistical test for RGF has to be made. When the response Y is binary, the test coincides with DeLong's test. In general, a statistical test based on U-statistics can be derived. In all cases, the higher the RGF, the higher the predictive accuracy, and the smaller the p-value, the more significant is the found RGF with respect to a given lower bound.

Let's exemplify the calculation with a numerical example that concerns a classification problem.

Consider the classifications obtained in Table 2-6 in Chapter 2, and suppose we remove a protected variable from the model, such as gender, race, or nationality.

From Table 5-1, the sum of Q'i-Qi (maximum possible perturbation) is equal to 1.31, whereas the sum of Q'i-QPi (actual perturbation effect) is equal to 0.89, leading to an RGF equal to 68%, a value of fairness that indicates attention.

Table 5-1. *Calculation of the RGF measure for a classification problem (change values)*

Individuals	Observed Response	Predicted Response	QFi	QFi'	Predicted Without Protected Variable	Ranks Without the Protected Variable	QPi	Q'i-Qi	Q'i-QPi
2	0	0.10	0.04	0.32	0.60	3	**0.19**	**0.28**	**0.13**
5	1	0.46	0.21	0.58	0.46	1	**0.34**	**0.37**	**0.24**
1	0	0.52	0.41	0.79	0.52	2	**0.50**	**0.38**	**0.29**
4	1	0.71	0.68	0.96	0.71	4	**0.73**	**0.28**	**0.23**
3	1	0.85			0.85	5	**1**		

We conclude this section with the Python code required to calculate the RGF, for any given machine learning application.

The function is called

check_fairness

It provides different functions related to the calculation of the Rank Graduation Fairness measure (RGF).

The RGF function can calculate the Rank Graduation Fairness (RGF) of a model, for classification (categorical response) or for prediction (continuous response).

More details on the Python implementation, including a statistical test for RGF, which provides a corresponding p-value, are provided in the appendix. What is presented for tabular data can be extended to all types of data, and it can cover generative AI as well. To this end, it is necessary to identify words or sentences that can be associated with a protected variable.

Scoring Rubric

Once the outlined measurement metrics for RGF have been completed, it's important to score the robustness of your model(s). Here's a simple scoring rubric for fairness.

Table 5-2. *A scoring rubric for fairness*

	Excellent	Good	Fair	Borderline	Poor
What is the fairness score of your model(s)?	75%–90%	70%–75%	60%–70%	40%–60%	0%–40%

Mitigation

Once you've determined the fairness scores of your models and applications, it's time to fix and reduce the biases and unfair results you detected.

Here are a few ways to resolve any fairness issues identified in your AI systems:

1. **Apply explainability methods**: Using the RGE metrics discussed in the previous chapter to fix a poor explainable score in an AI model is one sure way to resolve fairness issues as a more explainable model pinpoints areas where the model could be defaulting on fairness.

2. **Make changes to the datasets**: By resampling the dataset and adjusting the training data to represent more samples (oversampling) from underrepresented groups, fairness issues can be resolved as there'll be adequate representation of all groups in the dataset, which will balance it out.

3. **Leverage on synthetic data**: You can also improve the diversity of the dataset by creating synthetic data for the underrepresented groups in the dataset, based on the fairness results when tested.

In this chapter, we looked at fairness and its importance and how to measure fairness using the Rank Graduation Fairness (RGF) score. We looked at examples where the global Shapley method was applied to a logistic regression model and a random forest model to investigate racial discrimination in a loan dataset. We concluded this chapter with a few mitigation measures for poor fairness scores. In the following chapter, we'll look at privacy.

CHAPTER 6

Privacy

When working with AI applications and models, the privacy of individuals is of utmost importance. Privacy in machine learning aims to preserve the privacy of individuals represented in the datasets. It tends to employ and use different techniques to safeguard sensitive data during ML development and deployment. In this chapter, we'll look at privacy and how to measure privacy in ML models and algorithms.

To preserve privacy, the NIST AI RMF recommends an AI application should be as independent as possible from individual observations. In other words, omitting information about a single individual or about a given set of individuals should not have a relevant impact on the output. If this is the case, such observations can be removed without altering the AI output substantially. These considerations allow us to measure the degree of privacy and the degree of compliance with the privacy requirement, examining the dependence on the output on a given set of observations.[46, 47] This can be achieved with the SAFE-HAI framework proposed in this book.

The Rank Graduation Accuracy (RGA) measurement can be extended to a Rank Graduation Privacy (RGP) measurement, comparing the predictions obtained with a model based on all data with those obtained from a model that excludes one or more data points which needs to be "forgotten" – an important principle in machine learning where personal data should be erased based on different circumstances, such as when personal data is no longer needed, or individuals have requested for the removal of their personal data. The RGP is based on the difference between the observed cumulative predicted values calculated using all data minus the considered one.

T. Duke and P. Giudici, *Responsible AI in Practice*,
https://doi.org/10.1007/979-8-8688-1166-1_6

The RGP metrics can be obtained by considering the area between the dual Lorenz curve and the concordance curve, derived from the ranks of the predictions obtained without the data to be forgotten, and dividing it by its maximum possible value – the area between the dual Lorenz curve and the Lorenz curve.

To derive RGP, consider a set of N individuals corresponding to the observations in the rows of the training set. Let m be the set of individuals whose information is removed from the training set, for privacy purposes. Let us then denote the individuals corresponding to the observations in the rows of the test set with n.

To introduce a metric that can assess the compliance to privacy, we can follow four main steps, as follows:

> **Step 1**: A machine learning model is fitted on the whole training set, including all n observations.
>
> **Step 2**: The model fitted in step 1 is employed to obtain predicted values for the n observations in the test set.
>
> **Step 3**: m observations (with m < N) are removed from the training set, and the model is trained on the remaining N-m observations.
>
> **Step 4**: The model fitted at step 3 is employed to obtain predicted values for the n observations in the test set.

After these steps are taken, a new metric called Rank Graduation Privacy and denoted with RGP can be derived comparing the predicted values provided by the model fitted on the whole training set and computed on the test set, indicated with Yhat, and the predicted values provided by the model fitted on the training set without the m observations and computed on the test set, indicated with Yhat(-m). The Yhat values are reordered with respect to the non-decreasing ranks of the Yhat(-m) values, indicated with r^{-m}.

The Lorenz and dual Lorenz curves are built using the Yhat predicted values provided by the model fitted on the whole dataset (composed of n rows). The Lorenz curve is obtained cumulating such values in their ordered ranks, r. The dual Lorenz curve is derived cumulating the same values, but in reverse order. The concordance curve C is instead calculated using the Yhat predicted values reordered according to the ranks r^{-m} of the predicted values fitting the model on the reduced dataset, which does not contain the forgotten data (composed of N=n-m observations).

The Rank Graduation Privacy measure (RGP) can then be defined by

sum(Q'i-QPi)/sum(Q'i-Qi),

where Q'i is the dual Lorenz curve, Qi is the Lorenz curve obtained using the ranks r of the predictions obtained with all training data, and QPi is the concordance curve obtained using r^{-m}, the ranks of the predictions obtained excluding the data to be forgotten.

The RGP measure has some important properties, in particular:

1) It is normalized: RGP=0 in the case of irrelevant data; RGP=1 for the case of data with absolute importance; in general, 0<RGP<1.

2) To assess whether the RGP metric passes a set threshold indicating a minimum level of privacy, a statistical test for RGP can be derived, similar to the mentioned metrics in the previous chapters. The higher the RGP, the higher the value of the data, and the smaller the p-value, the more significant the found RGP is with respect to a given lower bound.

Let's exemplify the calculation with a numerical example. Consider the predictions obtained in Table 2-3 of Chapter 2, and suppose some data is removed from the model, resulting in the predictions in the sixth column in Table 6-1.

Table 6-1. *Calculation of the RGP measure for a predictive problem*

Individual	Observed Response	Predicted Response	Qi	Qi'	Predicted with Less Data	Predicted with Less Data Ranks	QPi	Q'i-Qi	Q'i-QPi
1	50	70	0.14	0.26	105	4	0.20	0.12	0.06
3	100	80	0.3	0.5	75	1	0.44	0.2	0.06
2	80	100	0.5	0.7	95	3	0.60	0.2	0.10
5	150	120	0.74	0.86	120	5	0.70	0.12	0.16
4	120	130			85	2			

From Table 6-1, it shows that the sum of the differences of Q'i-QPi is equal to 0.48, whereas the sum of Q'i-Qi is equal to 0.64. From this, the value of RGP is 0.48/0.64=0.75, indicating a reasonable level of privacy. Removing the data in the training sample for privacy purposes leads to a reduction in the model accuracy of about 25%.

This exercise can be repeated many times while considering different sets of training data to be forgotten, which can be specified by the user. An AI application will be deemed privacy preserving if the average RGP is significantly higher than a given value, determined subjectively by the AI developer. We can also evaluate whether the perturbation significantly deteriorates the prediction obtained with the original data with a statistical test, which is implemented in the Python code provided in the appendix of the book and described in Babei et al. (2024).[48]

More precisely, the function in the Python code is called

```
check_privacy
```

The function provides different functions related to the calculation of the RGP.

Let's illustrate its application in a real-life example based on a dataset that contains the salaries of the employees of a company. We will consider the assessment of privacy for both a classification and a prediction problem.

More precisely, we'll utilize the publicly available "Employee" dataset, well known to the statistical community, contained in the "stima" R package. The data derives from a study carried out on 473 employees of a bank and includes information on their gender, age, educational degree (in terms of years of education), employment category (custodial, clerical, or manager), job time in months since hire, total work experience (total job time in months since hire and from previous experiences), minority classification (i.e., whether of an ethnic minority), starting salary (in dollars), and current salary (in dollars).

A descriptive analysis of the data shows that most of the employees are less than 30 years old and that the majority of the employees have had 12 years of education, with a high degree of heterogeneity. The distribution of the number of months since the hiring date is also variable. The variable "previous experience" expressed in months shows a much higher variability than the previous variables, possibly because this variable is expressed in months.

For the data, we will consider both classification and prediction models. In the prediction model, the response variable is the value of the salary. In the classification model, the response variable is binary. It is "doubling the salary" obtained by assigning level 1 to the employees that achieve a salary growth rate (ratio of the current salary to the starting salary) greater or equal to two and a level 0 otherwise.

For more details, and to access the data and the Python code, please refer to the notebooks available on the GitHub repository indicated and described in Babei et al. (2024).

For the classification model, let's aim to predict whether an employee in the available bank data doubles the salary or not. The model could be employed to automatize (or semi-automatize) employee payroll progression, so it can be considered a "high-risk" application of AI, in the terminology of the EU AI Act.

Without loss of generality, we can model the classification problem with a random forest, RandomForestClassifier, as implemented in the sklearn.ensemble module of Python. More specifically, using the considered explanatory variables and the binary target variable, "doubling_salary" (y), the predicted values (yhat) are estimated for the test data (corresponding to the 30% of the whole dataset), employing the model fitted on the train dataset.

The RGP metric of the model can be calculated using the Privacy function in the Python check_privacy module.

To evaluate the privacy of a model considering the observations in the training data, it is necessary to evaluate the model while the observations of interest are removed from the model. For example, if individuals 10 and 12 ask to remove their data, RGP is equal to, respectively, 0.943662 (when we remove observation 10) and 0.929577 (when we remove observation 12).

The above model indicates that removing the 10th observation leads to an RGP equal to 0.94, higher than that obtained by removing the 12th observation. We can therefore conclude that the 12th observation is more valuable for the model.

To assess the compliance of the model with the privacy requirement, we can compare the values against a set threshold, when available. As for explainability and fairness, we can also assess whether the removal of one or more observations significantly changes the model.

For example, the RGP-based test for the tenth observation leads to a p-value equal to 0.003418. The p-value of the test is small, which means we'll reject the null hypothesis. Rejecting the null hypothesis means the output of the model significantly changes. This signifies that the model does not preserve privacy, according to the RGP metric. This result is expected, as the number of available observations in this database is relatively small (473 in total).

We can now consider the same data, viewing it as a response variable and considering the salary growth of the employees rather than its binary version (doubling of the salary). Our aim is to predict "salary growth," which represents the difference between two variables, "salary" and "starting salary." The resulting variable, the salary growth, is a continuous variable, and to predict it, we will consider a random forest model using the RandomForestRegressor function in the sklearn.ensemble package of Python. It's important to note that the choice of a random forest model is just one possibility that eases the interpretation of the results, as we use the same type of models for both classification and prediction.

We calculate, similar to the calculation in the classification problem, the RGP values when we remove the 10th and the 12th training observations. To this end, each observation is removed from the training data, and then the ranks of the predicted values estimated by the model including all the observations are compared with the ranks of the predictions estimated by the model excluding the selected observation, as previously described.

We obtain that RGP is equal to 0.995943 (removing the 10th observation) and to 0.996644 (removing the 12th observation).

Thus, removing the 10th observation leads to a high RGP, close to one, meaning that removing this observation from the model does not affect the predictions. A similar result also holds for the 12th observation. To confirm the results, we can apply the RGP-based statistical test for the 12th observation. The resulting p-value is equal to 1.65*10-08, which is a very small value. This leads us to reject the null hypothesis, which means that the ranks of the predictions by the reduced model are significantly different from those of the complete model, and the prediction model is not compliant to the privacy criterion, as it occurred for the classification model. Both models use very little data, making it difficult to be compliant with the privacy requirements.

Scoring Rubric

Once the outlined measurement metrics for RGP have been completed, it's important to score the privacy of your model(s). Here's a simple scoring rubric for privacy.

Table 6-2. *A scoring rubric for privacy*

	Excellent	Good	Fair	Borderline	Poor
What is the privacy score of your model(s)?	95%–100%	90%–95%	80%–90%	70%–80%	0%–70%

Mitigation

To ensure high levels of privacy of user data in your ML models, algorithms, and applications, you can adopt several ML privacy-preserving techniques, such as differential privacy, federated learning, homomorphic encryption, adversarial training, and secure multi-party computation (MPC).[49] Let's take a quick look at these techniques in more detail:

1. **Differential privacy**: This introduces a mechanism of introducing "noise" or perturbations to the training datasets, making it difficult for any attacker to identify personal identifiable information or any individual record. This enables the release of aggregate statistics in ML models without compromising or revealing the privacy of the underlying data.

2. **Federated learning**: Federated learning is a decentralized approach that enables multiple devices to collaborate on training a shared model without directly accessing their data. Each participant trains the model using their local data, and the model updates are combined to create a global model, without access to the personal data of each local dataset.

3. **Homomorphic encryption**: This employs a cryptographic encryption which doesn't require decryption of encrypted data before use and permits computation with encrypted data. This means that ML practitioners can build ML models trained and applied on encrypted data, which protects sensitive user data and information, ensuring they remain protected throughout the computation process.

4. **Adversarial training**: Adversarial training is an important ML privacy-preserving technique that is also used during ML safety tests. It involves training ML models to be robust against adversarial attacks and malicious acts. This is carried out by including adversarial examples in the training process, enabling the models to be more resilient and less susceptible to privacy and security breaches.

5. **Secure multi-party computation** (MPC): Secure MPC allows multiple parties to compute jointly over their inputs without revealing the individual inputs of the parties. This is a useful technique when you have several organizations or individuals working together on a particular ML task or project, but don't want to provide broad access to their individual data or share their sensitive data directly.

These methods have been introduced and employed by major tech companies, such as Apple, Google, and IBM.

To conclude on this chapter, we've looked at the importance of privacy, how to measure privacy, a rubric score for scoring the levels of privacy in your ML models, and different ML privacy-preserving techniques that can be utilized. In the next chapter, we'll look at sustainability.

CHAPTER 7

Sustainability

In the previous chapter, we looked at privacy and different ways to measure the levels of privacy in ML models and applications using the RGP score (Rank Graduation Privacy). In this chapter, we'll discuss sustainability and the requirements for an AI system to be "sustainable" in accordance with environmental, social, and governance (ESG) standards.

ESG can be defined as a set of standards used to screen potential investments in a company on the basis of how the company performs in environment (E), social (S), and governance (G).

ESG scores are very important for artificial intelligence applications, and this importance is reflected in the existing regulations. For example, the EU AI Act, while acknowledging that AI contributes to a wide range of environmental benefits "across the entire spectrum of industries and social activities," also states that "when a general purpose AI model poses a systemic risk, supervisors may consider the estimated energy consumption of training AI models." At a higher level, the United Nations AI Advisory Board states, at the beginning of its interim report (article 4), that "a key measure of our success is the extent to which AI technologies help achieve the Sustainable Development Goals (SDGs)."[50] Environmental, social, and governance factors are highly related to all 16 SDG goals so that, if AI applications take into account ESG factors, they may help to achieve SDG goals as well.

T. Duke and P. Giudici, *Responsible AI in Practice*,
https://doi.org/10.1007/979-8-8688-1166-1_7

Industrial sectors or verticals which have a high impact on the environment, such as energy and fashion, to name a few, may strongly benefit from the inclusion of ESG criteria in their key performance indicators. Better ESG scores can make companies in these sectors more attractive for investors and consumers, as they are perceived to be more sustainable in the long run.

From a financial perspective, contrary to more traditional financial indicators, the criteria for ESG aims at enhancing returns while simultaneously promoting positive social and environmental outcomes.[51] The incorporation of such standards in the investment decision-making process is defined as "sustainable finance," and it is meant to increase long-term investments in sustainable economic activities and projects.[52]

Despite the unstoppable growth ESG has experienced since the Great Recession in 2008, authorities have come up with little or no regulation on the matter, leaving rating agencies free to decide their own methodology for the assessment of a corporate's sustainability performance. In a similar context, the interpretation of ESG scores as provided by major financial platforms – such as Refinitiv or Bloomberg – becomes challenging, as the reasons behind the scores may not be clearly interpretable. Indeed, more and more firms engage in "greenwashing" behaviors, ultimately misleading stakeholders about their sustainable performance.[53, 54]

Though there is no unique classification yet, the features included in each of the three dimensions (E) (S) (G) could be described as follows:

> E, which stands for "environment," typically refers to a company's effort on climate change mitigation and adoption. It may account for a company's limitation on the usage of harmful pollutants and chemicals or to a company's active engagement in the reduction of greenhouse emission.[55] Among the three ESG items, environment (E) has always been the most popular one, both in terms of implementation on the firms' side and in terms of evaluation by rating agencies.

> S, which represents "social criteria," usually refers to the policies promoted by a company on the matter of inequality, inclusiveness, labor relation, and human rights issues. A company's business relationships alongside employees' health and safety and working conditions may be evaluated as well.
>
> G, representing "governance," may account for the presence of conflict of interest in the choice of board members, the percentage of female board members, or, in general, the percentage of gender diversity in a company.

In the next section, we will consider how to take into account ESG goals, in machine learning models, so as to improve the sustainable development not only of AI applications but also of our world at large.

Environmental Sustainability

Environmental sustainability is the first ESG factor and is related to the requirement that AI applications should not create harm for the environment.

The assessment of environmental sustainability is the most common ESG assessment, which has the most reliable and most sources of data. We explain how to measure environmental sustainability by means of a case study described in Agosto et al. (2023),[56] which looks at the relationship between ESG and the reputation of a company (rating). An investor company may employ AI models and would need to measure the environmental sustainability of AI systems by measuring the environmental sustainability of the company that produces, deploys, or uses an AI system.

Figure 7-1 describes the data we'll review as an example. This data is a sample of 1382 European companies and their financial reputation (credit rating). The figure describes the distribution of the companies in terms of credit ratings.

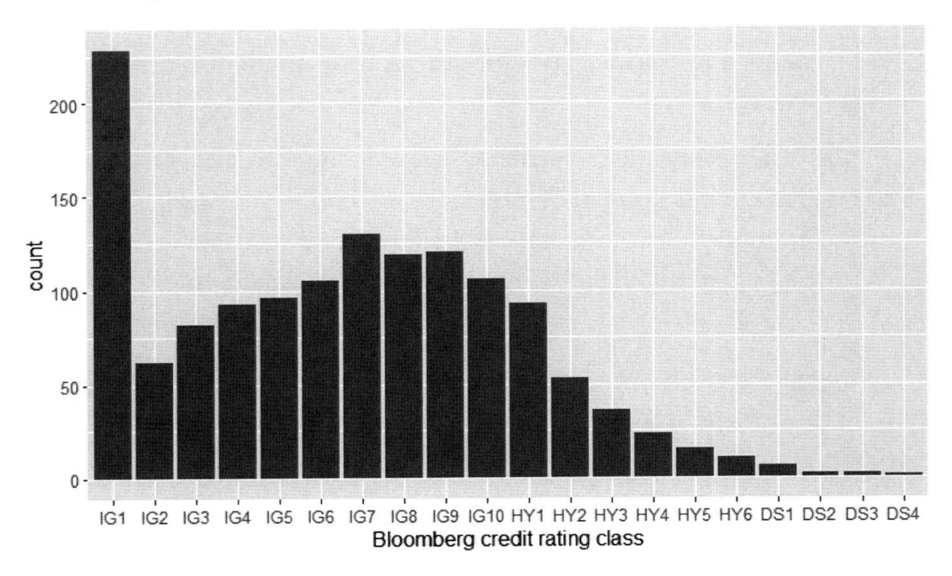

***Figure 7-1.** Distribution of the credit ratings for selected companies. Source: Agosto et al. (2023)*

Our goal is to build a model aimed at predicting the financial reputation of the company in question on the basis of the ESG score with a focus on the environmental score. On completion of this type of model, a powerful and useful tool could be utilized by the company to monitor and improve its reputation by varying the levels of environmental sustainability.

To explain the credit ratings, we consider using the overall ESG score of a company, as measured by different data providers. ESG scores produced by different providers are typically different, as they are built using different background variables and attribute different weights to the considered variables.

Agosto et al. (2023) introduced a Bayesian ensemble model which can produce a combined ESG score, weighting each individual score by means of a weight that is related to the predictive accuracy of the corresponding credit rating. More precisely, the combined score is obtained by attributing a weight to each available ESG score. This weight is a function of the likelihood of the observed counts of companies belonging to the different credit rating classes, under the alternative partitions generated by the ESG scores. The likelihood weights are expressed in-sample predictive performance and are obtained through the application of Bayes' theorem, as shown in Agosto et al. (2023).

The model is based on the assumption that there is an effect of ESG scores on credit rating. However, the aim is not to build a model that employs ESG scores to improve credit rating predictive accuracy but rather to investigate the relative importance of each ESG data score.

To this end, it is assumed that each ESG score generates a partition of the companies in j clusters (such as the four quartiles of the ESG distribution) and that the rating of a company is the same for the companies belonging to the same cluster. By doing so, we have one partition of companies for each given ESG score (each one supplied by a different data provider).

More formally, assuming that the rating is obtained from the estimated "probability of default" of each company (PD) and that each ESG is associated with a partition that places each company in a cluster j, the estimated PD is obtained as follows:

PD = PD(cluster j under partition P)*Probability(partition P).

This equation summarizes the essence of the model. It proves that it is a safe and Responsible AI model as it's composed of several Responsible AI principles from sustainability, fairness, and explainability. It's a sustainable model, as it allows us to measure the impact of ESG factors on credit ratings. It is a fair model, as it averages the contribution of different ESG providers, compensating their differences, due to different objectives. It is an explainable model, as it is a linear combination of weights with posterior probabilities, which, although calculated in a nonlinear way, have a clearly interpretable meaning.

For the purpose of measuring accuracy, let's evaluate if the model is accurate using the available data – in other words, whether ESG factors have a predictive relevance for credit ratings, and if so, what are the relative weights of each ESG factor in the model.

Specifically, we will apply the Bayesian ensemble model to a sample of 1382 European companies where we have retrieved the following information:

- **The Morgan Stanley Score Index (MSCI) ESG Score**: A continuous variable ranging from 0 (lowest sustainability) to 10 (highest sustainability).
- **The Refinitiv ESG Score**: A continuous variable ranging from 0 to 100. As for the MSCI ESG score, higher values indicate better sustainability profiles.
- **The Standard and Poor's (S&P) Global ESG Rank**: A discrete variable defined as the total sustainability percentile rank, ranging from 0 (lowest sustainability) to 100 (highest sustainability).
- **The rating class assigned to the company based on the Bloomberg Issuer Default Risk model generated probability of default over the next one year**: An ordinal variable whose categories are in the sample range from IG1 (highest credit worthiness) to D4 (lowest credit worthiness). Specifically, classes from IG1 to IG10 identify investment grade bond issuers, while classes from HY to H6 and from D1 to D4 identify high yield and distressed bond issuers, respectively. Starting from the rating class information, we define a binary variable which is equal to 1 if the company belongs to a speculative (high yield or distressed) class, 0 if otherwise. This will be our target variable.

- **A set of 13 financial ratios**: Return on equity, return on asset, return on investment, short-term debt on one-year growth, total debt on one-year growth, free cash flow on one-year growth, free cash flow on five-year growth, EBITDA (earnings before interest, taxes, depreciation, and amortization) to interest expenses, long-term debt to total equity, quick ratio, capital expenditure ratio, financial leverage, and asset turnover, which should reflect company profitability, growth, and liquidity, together with the value of market capitalization, which serves as a dimensional indicator.

The distribution of sample companies among the credit rating classes has already been shown. It shows that, for the considered companies, the distribution of ratings is quite skewed to the right and that there is a large group of companies with very high ratings. Both aspects will make it more challenging to attain a good level of predictive accuracy.

Figure 7-2 shows the distribution of the ESG scores for the same companies.

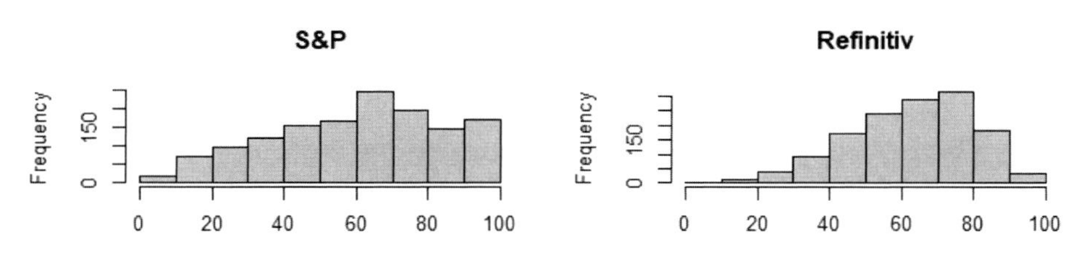

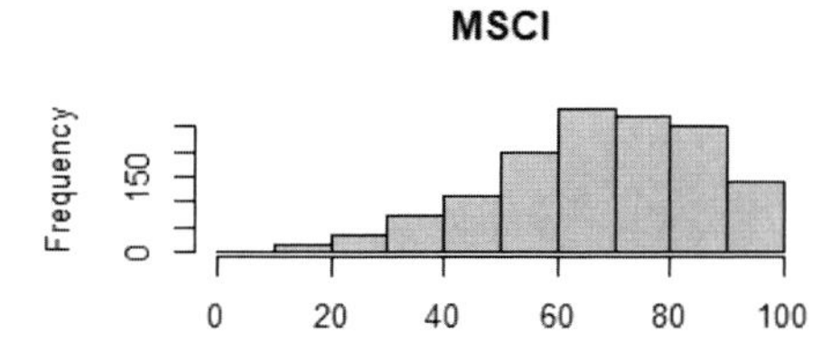

Figure 7-2. *ESG scores using the Bayesian ensemble model. Source: Agosto et al. (2023)*

In Figure 7-2, we can see that the distribution of the three ESG scores in the analyzed sample is left-skewed. In general, this shows that a good number of companies have shown a much worse ESG evaluation.

Calculating the correlations for the data in Figure 7-2, using the standard covariance/variance formula, shows that the correlation between all three pairs of ESG scores is quite low: 0.372 (MSCI-S&P), 0.383 (MSCI-Refinitiv), 0.692 (Refinitiv-S&P).

The application of the model to the data leads to the following weights: 0.36 (MSCI), 0.37 (Refinitiv), 0.27 (S&P). This means that to predict credit rating and the reputation of a firm, S&P ESG ratings are the least important, while the other two are more important. The difference between model weights decreases when we insert the balance sheet ratios in the analysis, but maintains the described order of importance.

We can consequently use the weights associated with the ESG scores estimated on the training sample (60% of the available observations) to predict the credit rating in the validation sample (40% of the available observations). The weights are then used to determine for each company and for each provider domain (Refinitiv, Standard and Poors, MSCI) the probability associated with each of the two considered rating categories: investment grade or speculative (high yield or distressed) class.

The results of the analysis show that the combined model has a predictive accuracy of about 61%, compared to the 59%, 55%, and 51% obtained by the single scores when examined separately. Therefore, we can conclude that the combined model slightly improves the predictions in a context in which predictive accuracy is difficult to obtain. More importantly, the model explains why an ESG score is better than another, by means of a set of explicit weights. For the considered data, Refinitiv and MSCI are more important than S&P. It can also be shown that the combined model is much superior than the others in the extreme classes (low rating classes). Similar conclusions are obtained when controlling for the financial variables.

Now we've discussed how to take environmental sustainability into account when building a machine learning model, in the next section we'll provide an example of how to embed social (S) and governance (G) sustainability factors in AI models. To this aim, we will consider gender diversity, which refers to both S and G.

Social and Governance Sustainability

In this section, we'll evaluate how to measure social and governance sustainability through a case study described in Bosone et al. (2022),[57] which concerns the relationship between "female presence" in companies, ESG, and financial performance.

The assessment of "female presence" by means of measurable indicators such as the number of female directors and the percentage of women in the Board could be categorized as measuring both social (S) and governance (G) factors. Though the problem of gender equality has been gaining considerable importance over the last decades, a number of related benefits are still unknown to many.

Bosone et al. (2022) addresses the question of whether the increase of female presence, which corresponds to an increase in the level of ESG factors, can also increase the financial sustainability of a company, as expressed by its rating, or probability of default. To this aim, and in line with the long period of time needed for ESG factors to manifest its effects, we decided to use as a response variable for the five-year default probability of firms. It is a continuous variable with values between zero and one. The choice of a five-year time period is in line with the fact that environmental, social, and governance scores, which include gender presence, are mostly used for investment decisions in long-term horizons.

To measure the effect of female presence in a company, several explanatory variables can be employed. The most widely used are the number of female employees; the existence of policies favoring inclusion or diversity; the existence of programs favoring work/life balance;

achievements in gender parity, including equal pay; and reports of controversies related to sexual harassment and discrimination. Social (S) and governance (G) scores are calculated taking into account all the previous indicators and, therefore, can be used as proxy measurements of gender equality.

Bosone et al. (2022) consider the S and G scores from the Bloomberg database for a relatively large sample of companies, not only aggregate ESG scores but also specific scores for each of the three dimensions E, S, and G. Trying to balance informativeness with data quality, they opted for the following measures of gender equality: the presence of equal opportunity policies (binary variable), the presence of health safety policies (binary variable), the percentage of women in the Board, the number of female directors and the number of female executives, the presence of female CEOs (binary variable), the existence of a gender pay gap breakout (binary variable), the fairness of the remuneration policies (binary variable).

To extend the width of the analysis, they also included the following: the presence of a Corporate Social Responsibility (CSR) sustainability committee (binary variable), the average and the total board compensation, and the aggregate ESG scores along with the specific scores for the social (S) and governance (G) dimensions.

A further binary variable for female-dominated (Fem_Dominated) and male-dominated (Male_Dominated) sectors was inserted, with the hope of obtaining some significant insights on the matter of gender pay gap across sectors.

Finally, the authors introduced some control variables at the firm level, namely: (I) the market capitalization, calculated by multiplying the total number of a company's outstanding shares by the current market price of a share; (II) the return on assets (ROA), a measure of how efficiently a company's management uses assets to generate earnings; (III) the return on equity (ROE), calculated by dividing net income by shareholders' equity, a measure of financial performance; (IV) the return on invested capital (ROIC), which expresses the capability of a company to extract value from its investments; (V) the weighted average cost of capital (WACC), in which each category of capital is proportionately weighted; (VI) the financial leverage, given by the ratio between total assets and total equity, which assesses the ability of a company to meet its financial obligations; (VII) the ratio between sales and revenues; (VIII) the ratio between return on capital (ROIC) and WACC, which can help to assess the performance of the company; and, finally, (IX) the credit rating, expressed by Bloomberg's analyst rating on a scale from 1 to 5 (1 represents the weakest value, a signal to sell a firm's shares; 5 represents the strongest value, a signal to buy a firm's shares).

The obtained database contains more than 15,000 cross-sectional data points from 2020 for 12 European countries: Austria, Belgium, Finland, France, Germany, Ireland, Italy, the Netherlands, Norway, Portugal, Spain, and Sweden. Companies from Austria, Belgium, Finland, Germany, Ireland, the Netherlands, Norway, and Sweden have been grouped under the label "North," whereas those from Italy, France, Portugal, and Spain have been grouped under "South." The number of companies for each block is comparable (329 Northern vs. 229 Southern companies), and the sample is well balanced.

The summary statistics show that the five years PD (probability of default) is on average slightly lower for Southern companies (7 basis points of difference: 4.43% against 4.78%), while the ESG scores of Northern companies are lower than the Southern ones (35 basis points of difference). Consistently, environmental (E), social (S), and governance (G) scores for Northern companies are on average lower than for Southern companies. It was also observed that the overall credit rating, as expressed by the probability of default, for Southern companies was higher (3.83% *vs.* 3.63%). This usually reflects the ESG and financial characteristics of a company.

Looking at the summary statistics for the gender equality-related indicators, we can come to the following conclusions: the number of female executives and the number of female directors are, on average, higher in Southern companies than in Northern ones. The same holds for the average percentage of women on Board (about 33% and 37% respectively). However, both the average and the total Board compensations are much higher in the North than in the South, in line with the average higher income. Within this framework, a further analysis of Bloomberg's database reveals that a fair remuneration policy is witnessed in 16 Northern companies out of 329, whereas only 7 Southern companies out of 228 are found to comply with it. Consistently, a gender pay gap is found to exist in more than half of Southern companies (128 over 228), whereas only 70 Northern companies out of 329 report it.

Since only 37% of women are currently employed in managerial positions across Europe, focusing only on female representation at top-level positions (e.g., number of women on Board, number of female executives) may provide a partial and incorrect view on the matter of gender equality. By contrast, the fairness of the remuneration policy and the gender pay gap at the company level can produce significant information.

A fair remuneration policy is supposed to comply with four aspects: minimum wage, fair wage, equal pay, and gender pay gap. In a company, the less fair the remuneration policy, the larger the pay gap is between top-level and low-level employees. As most women in the EU are typically employed in part-time, low-level positions, low level of income equality at the company level, this will inevitably cause the gender pay gap to widen. In line with this, larger gender pay gaps are associated with relatively unfair remuneration policies in Southern companies, while narrower gender pay gaps correspond to fairer policies in Northern companies. This is in line with common expectations, as Northern companies are on average embedded in a more egalitarian environment than Southern ones and are therefore more likely to implement policies favoring gender income equality.

To establish the relationship between gender presence and financial sustainability, Bosone et al. (2022) have implemented a logistic regression model. In the selected model, Bloomberg's five-year default probability is used as a response variable, while the country (North/South), the market capitalization (Mkt_Cap), ROE, ROIC, the ratio between ROC and WACC (ROC/WACC), the financial leverage (Fin_Lvrg), the analyst rating (Rating), the percentage of women on Board (Pct_Wom_BoD), the presence of CSR Sustainability Committee (CSR_Sust_Commitee), the total Board compensation (Tot_BoD_), the ESG scores, and individual score for social (S) and governance (G) are all used as explanatory variables.

They found a positive correlation between the five-year default probability and the binary variable country (which assigns 1 to Northern countries and 2 to Southern countries), in line with the summary statistics. A weak positive correlation (with a significance level of 10%) is found between the total ESG scores and the five-year default probability, whereas social (S) and governance (G) have a significant impact on the five-year probability (with significant levels of 5% and 1%, respectively).

More precisely, an increase by 10% in either social (S) or governance (G) scores leads to a decrease of the default probability by about 0.5 and 0.9%, respectively. These effects are counterbalanced by the opposite effect of the total ESG score, whose 10% increase leads to an increase of the default probability of about 0.7%. However, summing up the three linear coefficients, we see that the overall effect is negative, with a decrease in the default probability of about 0.7% when all of E, S, and G increase by 10%. If we compare the decrease in PD with the average PD of about 4.30, as seen in the previous section, we roughly obtain a 16% decrease of the PD, implied by ESG factors. This result demonstrates that by enhancing the share of ESG investment in social (S) and governance(G), firms can reduce the probability of default. Similar results are also evidenced by similar studies.[58, 59]

The authors also show that the percentage of women on Board is negatively correlated with the default probability; thus, the more women there are on the Board, the lower the risk of default. More women in the Board increase dialogue among board members, improve the quality of decision-making process, and favor the implementation of innovative and competitive business strategies, with a positive effect on corporate outcomes.

The obtained empirical evidence supports the validity of the research assumption: a higher presence of women in the Board and, more generally, higher social and governance scores decrease the probability of default of a firm, improving financial sustainability. This conclusion is reinforced by a found positive and significant correlation between the number of female directors and the presence of CSR committees, in line with the extant literature. This favors sustainability, not only from a financial viewpoint but also from a social and governance perspective.

We can thus conclude that a higher level of female presence is likely to improve a firm's performance both in financial and sustainable terms. More generally, environment (E), social (S), and governance (G) indicators can be used to assess the sustainability of a company and of a company that produces, deploys, or uses artificial intelligence applications.

Model Sustainability

The Rank Graduation Explainability (RGE) measure can be employed to assess the relevance of ESG factors, thereby measuring model sustainability. We can consider E, S, and G scores as explanatory variables and calculate their explainability as we did in Chapter 4.

Mathematically, we will obtain a Rank Graduation Sustainability measure (RGS), defined by

sum(Q'i-QPi)/sum(Q'i-QFi),

where Q'i is the dual Lorenz curve, QFi is the Lorenz curve obtained using the ranks of the predictions obtained with a full model, and QPi is the concordance curve obtained using the ranks of the predictions obtained excluding the variable under measurement which, in our case, is one of E, S, or G, or all together.

The RGS measure will have the same properties as RGE. We'll therefore forgo the example and the description of the Python code as it is the same as for RGE, described in Chapter 4.

Scoring Rubric

Once the outlined measurement metrics for RGS have been completed, it's important to score the sustainability of your model(s). Here's a simple scoring rubric for sustainability.

Table 7-1. *A scoring rubric for sustainability*

	Excellent	Good	Fair	Borderline	Poor
What is the sustainability score of your model(s)?	75%–90%	70%–75%	60%–70%	40%–60%	0%–40%

Mitigation

This chapter has shown the critical dependence to ensure ML systems are sustainable using the ESG factors. If you discover your systems are lacking in ESG factors and have a poor sustainability score, it's important you reduce the risks of your models. Reducing these risks will still adhere to the ESG principles of environmental, social, and governance.

Social and Governance

The sustainability of an ML model is highly dependent on its ability to have a positive impact on society.[60] In some areas, such as healthcare, AI is helping to provide more accurate medical diagnosis, driving drug discovery, assisting with higher-precision surgeries, developing personalized treatment plans, and so on. On the other hand however, AI still challenges and negatively impacts social cohesion and society through the various risks associated with it, such as bias and discrimination, hallucinations, inaccurate and incorrect outputs, stereotypes, representational harms, lack of transparency, privacy violations, human rights violations, threats to democracy, energy consumption, and security threats. To navigate these issues, aligning to the various recommendations in this book will help to resolve most of the challenges AI systems pose to society. Also, ensuring datasets are diverse and inclusive, and ML development/design teams are made up of diverse groups of people, will assist with this issue.

Economic and Environmental

Tackling the economic aspects of ML algorithms and systems, it's important that its benefits outweigh its cost. With the recent surge of generative AI, experts and business executives are beginning to question the value of generative AI as it demands high investment running into millions of dollars, with very little turnover, profit, or return on investment (ROI). While it might be early days to conclude on the overall value of generative AI to businesses given it's only been around for roughly two years, a few businesses have turned off their generative AI pilots and campaigns as they haven't seen value for money. For AI systems to be sustainable and economically viable, it's crucial they return value for money.

The second area for consideration is the impending job losses that AI has introduced. While AI should create new job roles, it's already started replacing human jobs with its various capabilities. Although the forecasted number of impacted jobs by AI might be quite speculative and yet to be proven, it's important efforts are made by organizations to ensure new job opportunities are created and workers have the necessary skills to navigate this disruptive technical revolution. To ensure AI systems achieve economic sustainability, it's important AI design includes the augmentation of human capabilities and not the replacement of them.

Finally, AI could play a major role in mitigating the impact of human activities on the environment. This is currently being carried out by many nonprofit and environmental organizations. However, the negative impact AI systems have on the environment is too high to ignore, largely due to the high amount of computing resources required to run these machines. From very large data centers to large GPUs (Graphics Processing Units) to support the billions of parameters needed for AI – particularly large-scale or foundation models – the energy consumption derived from these systems is quite high and concerning. While the potential for AI to massively improve lives and drive business value for organizations exists, we need to ensure environmental sustainability is also achieved. Reducing the carbon footprint of AI machines will lead to the promotion of more energy-efficient hardware and more greener algorithms. Work is being done in this area by various AI start-ups looking for alternative ways to develop AI chips and GPUs. Building smaller and more efficient AI models could also help to combat this issue.

To conclude, in this chapter we've looked at sustainability as a whole and its relation to ML models and applications. We also discussed the ESG factors and how to measure specific ESG values against countries and organizations with a focus on European countries and the finance sector. We introduced the RGS (Rank Graduation Sustainability) metric, which can be computed with the same methodology as RGE (Rank Graduation Explainability) score. We concluded the chapter with ways to reduce sustainability risks identified in AI systems.

In the next chapter, we'll discuss human-centered AI, which refers to the "HAI" section of the SAFE-HAI framework discussed in this book.

CHAPTER 8

Human-Centered AI

In the preceding chapter, we looked at an important and not very popular Responsible AI component, sustainability, using the ESG factors. In this chapter, we'll discuss another important topic in Responsible AI, known as human-centered AI or HCAI.

Artificial intelligence systems are tools that should be utilized by humans and should be subject to monitoring, control, and intervention by humans. This is where the term "human in the loop" or "human oversight" comes into play.

The European AI Act postulates that "high-risk" AI systems should be designed so that they can be effectively overseen by natural persons during the period in which they are in use. Human oversight shall then aim to prevent or minimize the risks to health, safety, and fundamental rights that may emerge when a high-risk system is being utilized.

Human oversight is also at the core of the American NIST risk management framework, where AI risk management is a key component for responsible development and the use of AI systems. The framework postulates that "AI risk management can drive responsible uses and practices by prompting organizations and their internal teams who design, develop, and deploy AI to think more critically about context and potential or unexpected negative and positive impacts." This establishes a direct correspondence between human oversight and AI risk management.

T. Duke and P. Giudici, *Responsible AI in Practice*,
https://doi.org/10.1007/979-8-8688-1166-1_8

In this chapter, we'll show how an AI model can be evaluated, assessed, and possibly modified to prevent and mitigate the risks associated with its use drawing on the EU AI Act, the NIST risk management framework, and other international frameworks aimed to promote safe and trustworthy AI. The employment of the SAFE metrics, along with the human oversight described in this chapter, leads to our proposed SAFE-HAI framework: a framework for sustainable, accurate, fair, and explainable artificial intelligence applications, controlled by humans.

To explain how to do human oversight in practice, we will utilize a real use case that concerns financial predictions, taken from Giudici and Raffinetti (2023).[61] The case study will be described in three main steps, which could be carried out without loss of generality, for any AI system:

1. The first step concerns adapting the SAFE-HAI model to the output from an AI system to derive compliance metrics.

2. The second step relates to the assessment of the results obtained from the SAFE-HAI model in terms of risk management.

3. The third step refers to the implementation of actions to improve the model output, including the construction of a new AI application which can improve the compliance of the available model.

A further use case of human oversight will be provided in Chapter 10, in a case study.

Evaluating AI

We consider, without loss of generality, the data described in Giudici and Abu-Hashish (2021),[62] which concerns a form of personal investment that is increasingly being used around the world due to its extensive availability and ease of access – the bitcoin. The data will be used to assess whether a machine learning model which is employed to predict bitcoin prices and therefore suggest investments or disinvestments is aligned with the SAFE-HAI model. To this aim, the main goal of our analysis will be to assess if the considered machine learning model is SAFE, according to our framework, so that it can be safely used by consumers and investors to determine whether and how to invest or disinvest in bitcoin.

Without loss of generality, we focus on the bitcoin prices from the Coinbase exchange as the target variable to be predicted. The time series for Oil, Gold, and SP500 prices are taken into account as candidate financial explanatory variables. In line with Giudici and Abu-Hashish (2021), the exchange rates USD/Yuan and USD/Eur are also included as possible further explanatory variables. The time series covers the time period between May 18, 2016, and April 30, 2018. Data from 2018 will be used as a test set and the others as the training set.

We are going to assess an AI application that can predict whether bitcoin prices go up or down the next day (a binary response) based on a machine learning model that is a feed forward neural network with one hidden layer and five hidden states. Before applying the neural network model, we'll transform all price series into their percentage returns. This is because returns are scale free and the corresponding series are stationary.

We will assess the machine learning model with the SAFE metrics proposed in the previous chapters. The same metrics could be applied to any other possible model, as the metrics are fully agnostic and require knowledge of the model output, but not model-specific information.

In line with our proposed SAFE-HAI model, we will apply the RGA metric to evaluate accuracy, the RGR metric to evaluate sustainability, the RGE metric to evaluate explainability, and, finally, the RGF metric to evaluate fairness.

The application of the RGA score to the output from the model gives a value of 55%: a low performance, which is expected as it is difficult to predict the price of bitcoin, a rather speculative asset, using classical financial assets.

The application of the RGR score gives a value of 81%: a relatively high value, indicating a relatively high robustness of the model.

The application of the RGE score indicates that the most important explanatory variable is Gold, with an RGE of about 17%, followed by Oil (6%), USD/Yuan (3%), S&P (3%), and USD/Eur (2%).

To assess fairness, we have introduced a "protected" variable which indicates the trading volumes of a day, split into high trading (traded quantities higher than the mean) and low trading (traded quantities lower than the mean). The obtained value for RGF is equal to 71%, indicating low fairness.

The application of the proposed statistical tests indicates that accuracy is not significantly higher than that of a random model, whereas robustness is; Gold is the only significant explanatory variable; the model is not significantly fair.

Assessing AI

Let's now assess the results – the application of the SAFE-HAI model indicates that the built AI application aimed at predicting whether bitcoin prices increase or decrease does not seem to be compliant with the accuracy and to the fairness requirements (assuming trading volumes as the protected variable). It is instead compliant in terms of robustness and, to some extent, in terms of explainability. The latter as one explanatory variable (gold) is significant, leading to an easy to interpret model which relates bitcoin prices to the price of gold. In fact, the assessment of the model seems to confirm the story of bitcoin being "the digital gold."

To entertain mitigation actions, the values of the compliance metrics and the related results of the statistical tests must be translated into risk management results.

The risk of harm caused by an AI application has two components: the probability that a harm occurs and its severity, that is, the quantification of the actual harm, when it occurs.

The probability component can be measured mapping the values of the SAFE-HAI metrics to levels of probability that complement the metric values. For example, an RGA of 55% can be mapped to a probability of 45%. An RGR of 81% can be mapped to a probability of 19%. An RGF of 71% can be mapped to a probability of 29%. And, finally, the sum of the RGE values equal to 31% can be mapped to a probability equal to 69%. These are the main results of our assessment and are at the basis of our proposed human oversight procedure. They indicate that the machine learning model under evaluation has a very high probability of not being explainable (=69%) and has a high probability of not being accurate (=45%). But, on the other hand, it has a low probability of not being fair (=29%) and of not being robust (=19%).

The severity component is more difficult to assess, and its mapping requires the consideration of the possible harm types involved with that incident and the stakeholders which may be impacted. Measuring the severity requires an appropriate taxonomy of harm types and stakeholders, as recently proposed by Abercrombie et al. (2024).[63]

The task of assessing the severity can be simplified using ordinal levels, such as "low," "medium," and "high." In the context of the machine learning model employed to fix bitcoin prices that we are examining here, the impact of a low accuracy is certainly high, as bad predictions may cause high financial losses to the investors, whereas the impact of a low explainability is low, as bitcoin prices are known to be not transparent. On the other hand, the impact of a low robustness will be high, and the impact of low fairness will be limited, as bitcoin is a technical tool, unlikely to be associated with specific population groups. Putting together the assessed probability and severity, and recalling that the risk of harm can be obtained as the product (or as the convolution) between the probability and the severity, we can conclude that the most important risk of the considered AI application is that of accuracy: a classical "model" risk. The conclusion of the oversight assessment of the considered model that predicts bitcoin prices is that the model is not good, particularly in terms of accuracy, and that it should, therefore, be replaced by a better model.

Improving AI

We have found that the model should be improved in terms of accuracy. To this aim, we have implemented alternative neural network models that take into account the time dependency present in time series, such as the series of bitcoin prices analyzed in the case study of this chapter, to predict future prices.

Bitcoin is a speculative asset, and, for that reason, buy/sell decisions about bitcoin are usually based on past prices of the same, rather than on the behavior of other assets, as different types of investments.

The model considered in the previous section is a feed forward neural network that assumes that bitcoin prices of different days are independent of each other. Given the weak performance of the model, in terms of accuracy, we'll now consider three different types of neural networks that can take into account the time dependency embedded in the time series data to generate forecasts and scenarios.

Specifically, we will consider the autoregressive neural network autoregressive model (NNAR), long short-term memory (LSTM) networks, and gated recurrent unit (GRU) networks.

The structure of a NNAR model is described in Figure 8-1.

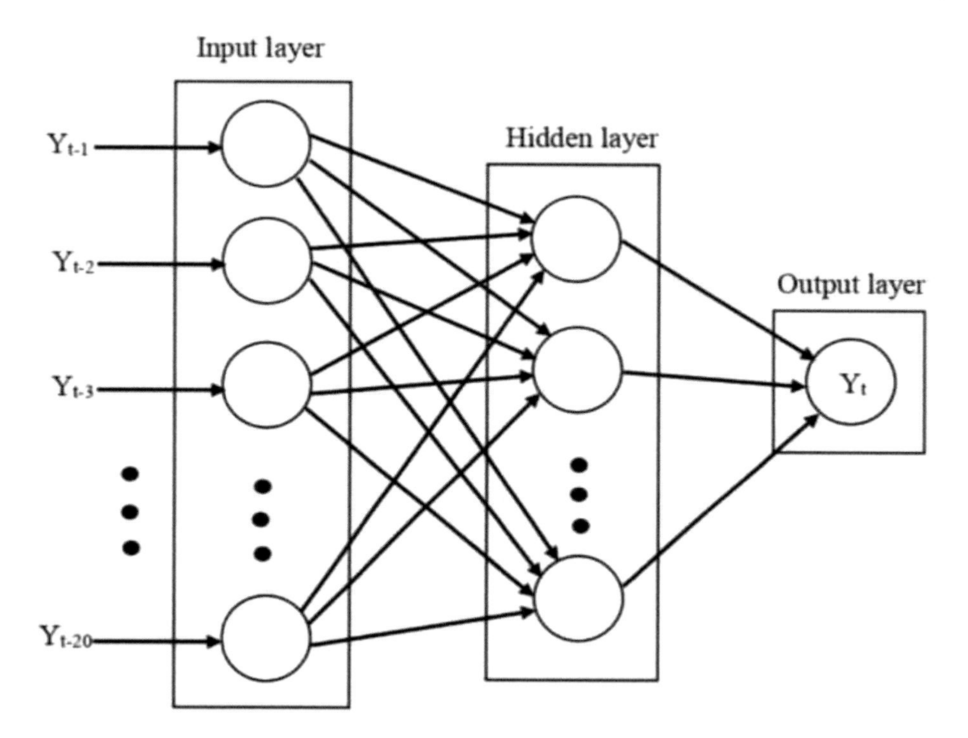

***Figure 8-1.** The structure of a neural network autoregressive (NNAR) model. Source: Giudici, Piergallini, Recchioni, and Raffinetti (2024)*

From Figure 8-1, note that the NNAR neural network in the input layer has p lagged values of the response variable, similarly as in the classic autoregressive (AR) models used in financial econometrics. Different from what occurs in the AR model, however, the input is not connected directly with the response, but is processed by an intermediate hidden layer, a computational unit which transforms the signals from the input before employing them to determine the output values.

It follows that a NNAR model is a network characterized by p input nodes, the lags of the response variable y(t-1), y(t-2), ... , y(t-p), and the neurons in the hidden layer. Now let's look at the next network we'd like to consider, long short-term memory networks (LSTM). Long short-term memory networks (LSTM) can be described by a sequence of recursive steps, in which each input x_t is processed as in Figure 8-2.

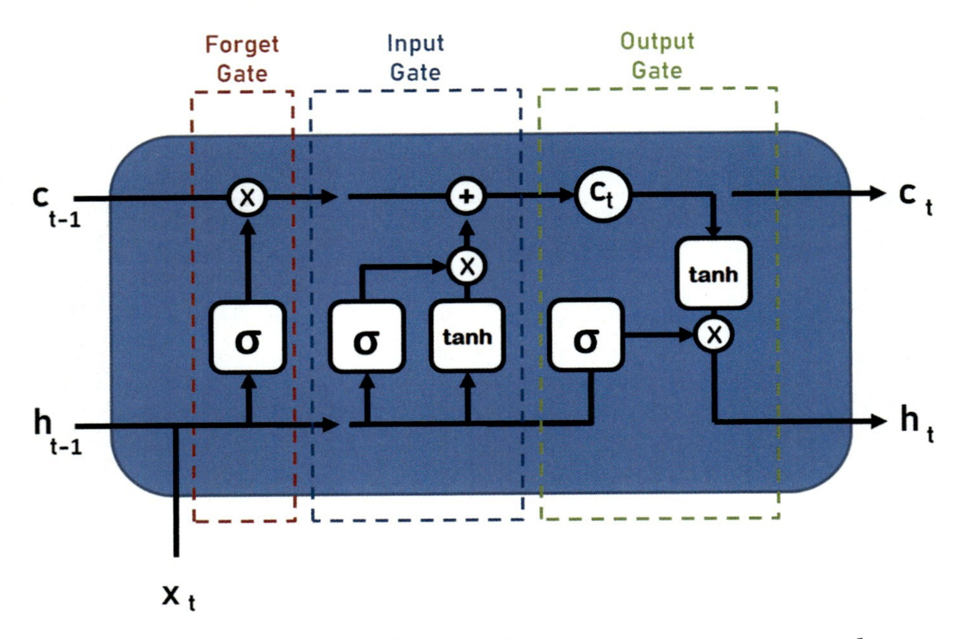

Figure 8-2. *The sequence of a long short-term memory network (LSTM). Source: Giudici, Piergallini, Recchioni, and Raffinetti (2024)*

From Figure 8-2, note that during the processing of each unit x_t, there are two flows of information represented by the two horizontal lines that enter and exit the LSTM: the cell state c_(t), known as long-term memory (LTM), and the hidden state h_(t), known as short-term memory (STM).

As an alternative to LSTM networks, we will also consider gated recurrent unit (GRU) neural networks, whose processing units have the structure as in Figure 8-3.

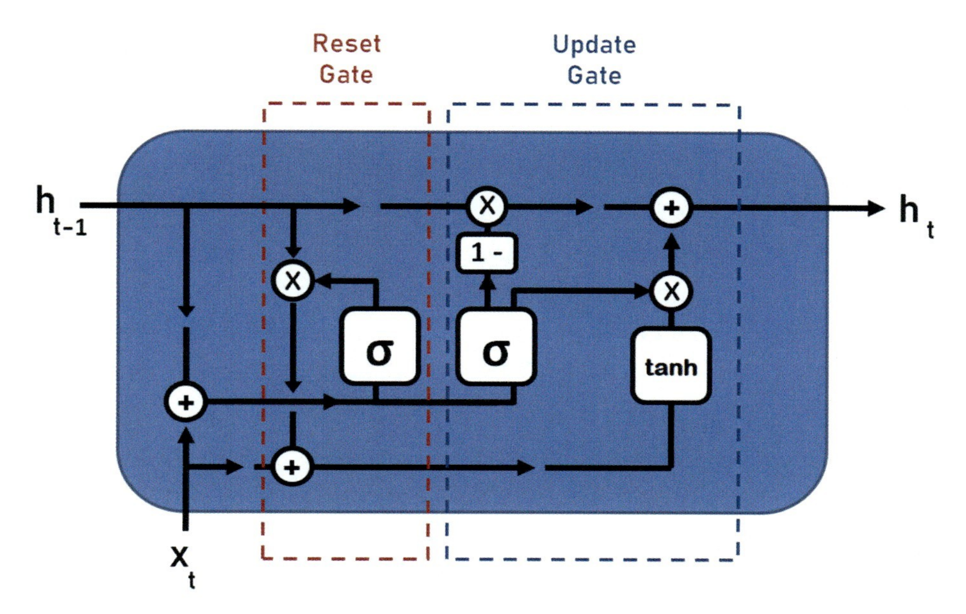

Figure 8-3. *An example of processing units of the gated recurrent unit (GRU). Source: Giudici, Piergallini, Recchioni, and Raffinetti (2024)*

A GRU network employs fewer calculations than the LSTM and is faster in the predictive phase.

More details on the presented NNAR, LSTM, and GRU machine learning models can be found in the paper by Giudici, Piergallini, Recchioni, and Raffinetti.[64]

We now apply the three described neural networks to the bitcoin price data previously described and compare their performance metrics against those of the time-independent neural network evaluated before.

In terms of accuracy, the RGA values for the NNAR, LSTM, and GRU networks are the following, respectively: RGA(NNAR) = 0.3718; RGA(LSTM) = 0.8186; RGA(GRU) = 0.8865. Recall that the RGA of the feed forward model of the previous section (our baseline) was equal to 55%.

The results indicate that, while the NNAR worsens accuracy with respect to the baseline, LSTM and GRU (and especially the latter) improve it considerably. From a closer viewpoint, it could be concluded that the NNAR includes too many input variables and leads to overfitting, while the LSTM and the GRU make a more parsimonious use of past bitcoin price data.

Moving to robustness, the baseline model has an RGR equal to 81%. It turns out that NNAR has RGR = 0.7157, a very poor value, whereas RGR(LSTM) = 0.9607 and RGR(GRU) = 0.9244. Both latter values indicate a robustness higher than the baseline, especially for LSTM. The conclusive explanation is that the NNAR has too many variables, leading to more unstable results, whereas GRU and LSTM are more parsimonious and, thus, more robust.

Moving to explainability, all new neural networks improve the total explainability, as expressed by the sum of the RGE of the predictors. The baseline model had a total RGE of about 31%, whereas RGE(NNAR) = 0.63, RGE(LSTM) = 0.56, and RGE(GRU) = 0.62. By choosing more predictors, NNAR does well in explainability, whereas the more parsimonious LSTM does a little worse.

How can we evaluate the results from the new networks? We already pointed out that lack of accuracy is, for this problem, the most important risk. We should therefore choose the most accurate model, which is the GRU model. It is important however to be aware that an LSTM model will perform slightly less in accuracy (and in explainability) but slightly better in robustness.

We can thus conclude the human oversight application suggesting that the AI application aimed at predicting bitcoin prices should change the machine learning model based on a feed forward network with a machine learning model based on a gated recurrent network.

Scoring Rubric

The scoring rubric for human-centered AI can be obtained by calculating the difference between the values of the metrics before and after an intervention action. For example, we can take the difference in the values of RGA, RGR, and RGE (and possibly of RGF) under different models and take the ratio between the difference and the original value. This can lead to a scoring rubric as in Table 8-1. The values in the table are an exemplification, and can easily be changed, in line with the desired model quality and related risk appetite.

Table 8-1. *A scoring rubric for human-centered AI*

	Excellent	Good	Fair	Borderline	Poor
What is the obtained improvement using the new model instead of the baseline?	50%–100%	30%–50%	20%–30%	10%–20%	0%–10%

Mitigation

Now that you have a thorough understanding of how to measure the value of human oversight, it's important you work on improving your model if it comes up with a low score.

Depending on the issue, there are various ways to improve a model, as we have seen. Most of these involve increasing the available data and/or changing the underlying machine learning model and algorithm.

In this chapter, we have considered how the initial model, based on a simple neural network, could be improved especially with regard to accuracy, using a more complex recurrent neural network. We also looked at human-centered AI (HCAI) and its importance in evaluating AI models, using a bitcoin analysis as an example. We discussed ways to ensure models can be properly assessed using HCAI. In the next chapter, we'll wrap up the different Responsible AI principles we've discussed in this book by reviewing governance processes and various harms associated with AI technologies.

PART IV

Governance and Case Study

CHAPTER 9

Governance Processes

In the previous chapter, we looked at human-centered AI (HCAI) with real examples using accuracy on a neural network and the importance of HCAI in AI development. We used bitcoin as an example. In this chapter, we'll deep dive into AI governance and how it plays a major role in the entire AI development process and life cycle.

In Chapter 1, we reviewed different AI governing policies and discussed the various components and how they relate to the SAFE-HAI framework. Expanding on the existing governing policies for AI, it's important that AI adopters, developers, and users have a thorough understanding of AI governance and create AI governance processes in their organizations.

AI governance refers to the structures and processes put in place to oversee an organization's AI strategy, management, and operations. With the ever-increasing risks identified in AI systems, setting up an AI governance structure is fundamental to ensuring successful AI development and use.

To understand how to properly govern AI, it's paramount to gain insight on the various risks and harms associated with the technology as well as the different stakeholders it affects. We referred to some AI risks in the first chapter. In this chapter, we'll deep dive into general risks identified in open source, general-purpose, and foundational models.

T. Duke and P. Giudici, *Responsible AI in Practice*,
https://doi.org/10.1007/979-8-8688-1166-1_9

Risks of AI Models and Applications

There are several risks and harms associated with AI models and applications, regardless of AI type or domain, in other words, whether it's an open source or a closed model. Most of the risks across the various types of AI systems overlap and have several similarities. NIST's risk management framework, referred to numerous times in this book, categorizes AI's harms across three areas:

1. **Harm to people**: This affects individuals, groups, and society. It occurs when an individual's civil liberties, human rights, or psychological and physical safety are adversely impacted by an AI system. Groups or communities can also be negatively impacted when biases and discrimination are portrayed as a result of AI technologies. Societal harm is also considered particularly when democratic participation is repressed or influenced by AI systems.

2. **Harm to an organization or enterprise**: An organization or enterprise is considered to have been impacted by an AI system when technical systems and business operations are affected by AI, such as cyber attacks, security breaches, monetary loss, or reputational harms.

3. **Harm to a system**: This occurs when a system of organizations is affected by an AI technology, such as a financial system or a global supply chain or an environmental ecosystem, which are not robust enough to withstand adverse AI impact.[65]

Let's deep dive into some of the various AI risks that pertain to the mentioned categories. Some of these risks include the following:

1. **Disinformation and election interference**: This refers to instances where several outputs from AI models contain incorrect information and are able to convince users of its credibility, even though responses are in error. This is quite common in LLMs. Due to this, a lot of focus is placed on social media sites which tend to have a proliferation of disinformation, and content moderation is necessary.

2. **Biorisk**: This is a foreseen and not yet proven risk particularly pertaining to open source foundation models, where several studies have claimed that users can receive instructions on how to create bioweapons. These studies are yet to be proven as some of the instructions observed from these models and applications have been quite incorrect and misleading. However, biorisk is considered a high risk in AI systems.[66]

3. **Spear-phishing scams**: Generative AI models are known to have the ability to generate spear-phishing emails. These are email scams that appear legitimate and target specific individuals or groups, tricking recipients to share sensitive information, download malware, or send money to the scammer.[67] The main point of concern is the ability to affect downstream tasks and applications without any safeguards or guardrails. However, this could be considered a low risk as most operating systems and browsers have high security against malware like these, and there's a high probability phishing emails may never get to the recipient in the first place.

4. **Voice-cloning scams**: We've seen an increase in voice-cloning scams in recent times, where malicious actors impersonate a person's friend or family and convince them to send money to their bank accounts. These impersonations use AI tools that are able to clone a person's voice using a short audio clip. These scams have also been used to spread disinformation, particularly in war-ravaged countries.

5. **Nonconsensual intimate imagery (NCII) and child sexual abuse materials (CSAM)**: Particularly associated with open source text-to-image models, NCII and CSAM images can be easily generated using these systems, including nonconsensual deepfakes. These go against several regulations and pose a huge risk to women and children.

6. **Authoritarian or corporate surveillance**: Authoritarian or Corporate Surveillance is commonly used across law enforcement agencies/ organizations and corporate bodies. In most cases, surveilled entities are not made aware of any surveillance measures and, even if aware, have no means to challenge the results of these systems. This has resulted in violation of human rights across several countries in various ways, such as false arrests, false penalties on welfare fraud, etc.

7. **Cyber attacks**: Most of the mentioned scams which use AI fall under the category of cyber attacks and have the potential to invade people's privacy and data, causing harm to individuals, communities, and organizations.

Governance Processes

To address and potentially govern the development and deployment of AI, in order to reduce the various risks associated with the technology, it's important to have a governance structure in place. Before making any informed decisions and policies, a review of the various stakeholders and users of AI is essential (apart from AI developers). These stakeholders are considered users, civil society, and policymakers who can make meaningful contributions to AI development, deployment, and use:

- **Users**: Users, many times, have a wider knowledge and understanding of AI applications as they tend to use them on a daily basis, even more than the developers themselves. They usually have various satisfactory levels with these applications and systems, pain points, and desired capabilities which could be considered as feature requests. Engaging with the users of these technologies is an important way to form governance policies.
- **Civil society**: AI education and accurate information about AI technologies could better inform civil society and the general public to contribute meaningfully to its development and deployment through standards, processes, research, policies, and so on.
- **Policymakers**: Policymakers play a major role in AI regulation either locally or internationally. Processes such as auditing, red-teaming, and safety evaluations can help guide policymakers on developing appropriate governing policies for the development, deployment, and use of AI.

To properly govern AI and develop the right processes and policies, adequate information about AI systems is required. This is illustrated in Table 9-1.

***Table 9-1.** Required information for appropriate governance policies*

Information Category	Requirements
Capability of the model	It's important to understand what the model is capable of. What tasks was it trained on? What datasets does it comprise? What are its error rates? False positives and false negatives? This provides insight on the potential of the model for misuse.
Controllability of the model	Insights on the intentional/unintentional behavior of the model are important. For example, does it reliably act according to the developer and user intentions? In cases of outputs, does it produce toxic content, for example? Answers to these will provide guidance on the ability for the model to cause unintended harm.
Model impact	Having a thorough understanding of the impact of the model on users, groups, and society at large is crucial. Does it produce representational harms, spread disinformation, produce incorrect/false content? To what extent will it potentially displace workers or replace jobs?
New and emerging information	Model developers might misrepresent their knowledge on the model or its impact for various reasons. It's important to verify developer claims through relevant information, testing, and risk assessments.
Uncovering new information	Most developers tend to test for accuracy and privacy only and fail to run safety and Responsible AI tests for their models, such as testing for robustness, security, fairness, sustainability, and data ethics. This could lead to a failure to identify relevant issues within their models, leading to further harms emanating from AI systems. Lack of scrutiny and understanding of AI systems can lead to various challenges that AI developers are unlikely to fix or be aware of.

Scrutiny, testing, investigation, and reporting of identified risks and harms help with the creation and development of governance processes for any AI domain.

To establish good governance practices, after a thorough risk assessment and ongoing reviews of the AI systems in an organization have taken place, the following need to be carried out as a governance structure:

1. **Define an AI policy**: Working with the legal team and other relevant stakeholders in your organization, ensure you have a clear AI policy in place. This goes hand in hand with AI principles which form the overarching guidelines and positioning of your organization's AI strategy. This policy should be publicized among the developer and engineering teams in the organization and other relevant teams. It should be clear, concise, understandable, and easily accessible.

2. **Establish an AI governance team**: Ensuring there is an AI governance team/working group that oversees and reviews AI practices in your organization is quite important to successful AI development and deployment. The governance team will set up and manage processes to ensure safe and Responsible AI design, development, and deployment. Establishing a multidisciplinary governance team made up of AI researchers, engineers, product, and legal teams will lead to the successful development of safe and trustworthy AI.

3. **Regular reviews and risk assessments of AI systems**: Conducting regular reviews and risk assessments of AI systems and products/applications in your organization is a critical part

of AI governance. Reviews and assessments should include the evaluation of datasets, models, and applications that use ML within the organization. These should be carried out on an ongoing basis with clear processes, escalation paths, and solutions to resolve and fix issues once identified.

4. **Inclusion of stakeholder engagement**: The involvement of various stakeholders, such as users, members of civil society, and policymakers, forms another key piece in AI governance. Regular meetings with these stakeholders will help inform the design of AI systems and how they are impacted. This will also help AI developers and organizations gain insights on user needs and how to better design/adjust the AI systems to be more suitable for users, society, and governmental organizations.

5. **Incorporation of user feedback**: Feedback from regular meetings with stakeholders should be incorporated into the AI systems, ensuring solutions are in place to support safe deployment and results. Feedback mechanisms need to be included in AI applications, such as a "thumbs up" or "thumbs down" button, where users can provide feedback on the accuracy of the application's responses. Processes to fine-tune and amend datasets, models, and applications should be set in place on a regular basis.

The risks and harms associated with AI technologies are too grave to ignore. Adopting the right AI governance process for your organization and measuring the risks using the framework introduced in this book will help to reduce some of these harms and solve any identified issues.

In this chapter, we had a purview of some of the risks identified in AI technologies, the various stakeholders involved in its engagement, and ways to carry out governance measures in your organization. In the next chapter, which is the final chapter of this book, we'll have an in-depth review of a case study which incorporates the SAFE-HAI framework in an application that relates to credit lending, considered a high-risk application of AI.

CHAPTER 10

Case Study

(Written in collaboration with Golnoosh Babaei)

In the previous chapter, we carried out a deep dive on governance, a review of the risks and harms associated with AI systems, and AI governance processes in an organization. In this chapter, we'll provide a case study that involves both the application and evaluation of a high-risk application of AI – a related case study can be found in Babaei et al.[68] The application involves the decision to grant credit loans through a credit scoring model. For this case study, we made use of the SAFE AI (security, accuracy, fairness, and explainability) statistical metrics available in the "safeaipackage" toolbox, a Python framework that can measure the security, accuracy, fairness, and explainability of a model. Considering the advantages of these metrics, namely, consistency (all derived from a common underlying statistical methodology: the Lorenz curve), easy interpretability, model-agnostic (applicable to any machine learning (ML) methods), and fully reproducible, by means of the available Python code, we apply this framework to this case study to assess the accuracy, explainability, fairness, and robustness of the considered ML models.

The main objective of a credit scoring model is to determine the credit risks of potential customers, that is, the determination of the probability that the customer will not pay back the credit on time (probability of default). Credit risk, a continuous variable with values in [0,1], is linked to the measurement of the customer's creditworthiness, an ordinal variable known as rating. The bank's decision (or other credit providers such as

T. Duke and P. Giudici, *Responsible AI in Practice*,
https://doi.org/10.1007/979-8-8688-1166-1_10

fintech and bigtech) on how much credit to grant and what interest rate to apply usually depends on the credit risk and, in particular, on the rating assigned.

In this chapter, we will examine how to determine the probability of default for credit applicants based on a set of explanatory variables. In particular, we will learn how to

1. Evaluate the accuracy of the most important and well-known machine learning models: logistic regression, random forests, and neural networks to classify the observations.

2. Determine which variables contribute to the model and which do not by assessing the explainability of the models.

3. Determine whether the model is fair.

4. Assess the robustness of the classifiers.

We will also discuss how to compare the predictions obtained from a machine learning model with the actual true values and calculate their differences as a main measure of model accuracy. Following this, we will compare the model accuracy of logistic regression, random forests, and neural networks.

The dataset used is a dataset containing 6707 observations, called the "training set." The data is made up of a binary variable, which we will use as a response variable (Y), called "default" with values equal to zero when the customer is considered reliable and equal to one when the customer is considered unreliable (risk of default or non repayment). In addition to the binary variable, the data contains the values of six explanatory variables (X1, X2, X3, X4, X5, X6) measured before the response variable: the interest rate, the amount of installment, the customer's annual income (expressed on a logarithmic scale), the ratio of the customer's overall debt to their income (Dti), the credit score assigned to the customer by the American

FICO software company, and the average days of delay in payment of installments. Table 10-1 represents some randomly selected rows from this dataset to provide an overview of the data in this case study.

Table 10-1. *The considered data: ten observations selected at random*

	Interest_rate	Installment	Annual_income (Log)	Dti	FICO	Days_of_delay
1	0.12	829.10	11.35	19.48	737	5639.96
2	0.11	228.22	11.08	14.29	707	2760.00
3	0.14	366.86	10.37	11.63	682	4710.00
4	0.10	162.34	11.35	8.1	712	2699.96
5	0.14	102.92	11.30	14.97	667	4066.00
6	0.15	344.76	12.18	10.39	672	10474.00
7	0.13	257.70	11.14	0.21	722	4380.00
8	0.11	97.81	10.60	13.09	687	3450.04
9	0.16	351.58	10.82	19.18	692	1800.00
10	0.14	853.43	11.26	16.28	732	4740.00

In Table 10-2, we share the descriptive statistics of the variables which are potential predictors of the response. The response variable is binary and describes whether a customer defaults (does not repay the loan) or not. The percentage of defaults in the data is equal to 0.195, which is quite common for consumer credit datasets.

Table 10-2. *Descriptive statistics of the explanatory variables*

	Interest_ rate	Installment	Annual_ income (Log)	Dti	FICO	Days_of_ delay
std	0.026834	207.0713	0.614834	6.88397	37.97054	6.88397
min	0.06	15.67	7.55	0	612	0
25%	0.1	163.77	10.56	7.2125	682	7.2125
50%	0.12	268.95	10.93	12.665	707	12.665
75%	0.14	432.7625	11.29	17.95	737	17.95
max	0.22	940.14	14.53	29.96	827	29.96
count	9578	9578	9578	9578	9578	9578
mean	0.122796	319.0894	10.93234	12.60668	710.8463	12.60668

The objective of this analysis is to determine a statistical model that estimates the relationship between the variable Y and the explanatory variables (X1, X2, X3, X4, X5, X6), then using this model and based on this relationship, predicting future values of Y (for customers not included in the training data) on the basis of the relative values of (X1, X2, X3, X4, X5, X6).

The target response variable considered in this chapter is binary. It is therefore not possible to apply linear regression, as it could lead to predictions for Y with values outside the interval [0,1]. While Y values within the interval [0,1] can be interpreted as "probabilities of default," with 0 indicating the absolute reliability of the customer (certain non-default) and 1 an absolute unreliability of the customer (certain default), values outside the interval will not be legitimate, as the probability has values in [0,1]. Therefore, a different predictive model is needed, such as a logistic regression. The selected ML models for the case study in this chapter are as follows.

Logistic Regression Models

A logistic regression model is a classification model which, in the mentioned example, allows us to classify each customer as reliable or unreliable based on the estimated probability of default of the credit granted to each borrower.

Unlike the multiple regression model in which the X inputs are used to take the value of the response variable, in a logistic regression model the inputs are used to classify the response variable into one of the available categories – reliable and unreliable as an example. The given result can be used to take the condition of each potential future customer, whether reliable or not, and allows you to decide whether to give or not to give a loan.

A logistic regression model is described by the following equation:

Here, for (i=1,2,...,n):

$$log\left[\frac{\pi_i}{1-\pi_i}\right]=a+\beta_1 x_{i1}+\beta_2 x_{i2}+\ldots+\beta_k x_{ik} \,..$$

The left side of the equation above defines the logit function of the estimated probability, as the logarithm of the odds of an event, that is, the natural logarithm of the ratio between the probability of success and that of failure.

Once calculated, an estimated value can be obtained for each dichotomous observation, by introducing a threshold value above which label = 1 and below which label = 0.

Our goal is to find estimates for the β_k values 1, 2, ..., k that best approximate the true relationship between Y and the set of X variables we have chosen. In the case of the logistic regression model, the relationship between Y and X is not linear and does not have an analytical solution. The solution can, however, be approximate, with an iterative numerical approximation method that is present in most statistical software.

Application of the Logistic Regression Model

Consider a bank that intends to build a predictive model of the default of its customers to find the relationship between the probability of repayment and a set of six independent variables: interest rate (X1); amount of the installment (X2); annual income (in log) (X3); debt/income (X4); days late in payment (X6); FICO score, that is, a measure of the customer's creditworthiness (X5). The equation is

$$log\left[\frac{\pi_i}{1-\pi_i}\right] = a + b_1 x_{i1} + b_2 x_{i2} + \ldots + b_k x_{ik} \,..$$

$$= a + b1x1 + b2x2 + b3x3 + b4x4 + b5x5 + b6x6$$

The application of statistical software to our data led to the following result:

log(Odds(D)) = -1.8327 -0.1589 installment amount + 0.0570 annual income (in log) + 0.0681 debt/income + -0.0591 days late in payment - 0.9630 FICO+ 0.3008 Interest rate

In other words, the software output provided the following coefficient values:

a = -1.8327, b1 =-0.1589 , b2 = 0.0570 b3 = 0.0681 , b4 = -0.0591

b5 = -0.9630 b6= 0.3008

Banking analysts can interpret the coefficients as the change in the odds of default. For example, an increase in the amount of the installment will lead to a decrease in the probability of repayment (and, therefore, of the odds) as well as for the days of delay and for the FICO score. On the other hand, an increase in annual income, debt/income ratio, and interest rate will lead to an increase in the probability of default. In all cases, the effect of each variable is considered without prejudice to the others.

The value of the coefficients must be interpreted from an economic point of view. While the results on the installment, days late, FICO score, and annual income are intuitive, those on the debt-to-income ratio and interest rate are less obvious. The first can be explained by saying that more debt (perhaps from other banks) allows one to repay the loan better; the second, by saying that higher rates lead to a greater incentive to repay the loan on time.

Verification of the Significance of the Logistic Regression Model

How to evaluate the goodness of fit of a logistic regression model? First of all, we can evaluate the significance of the individual coefficients by relating the estimate of each coefficient to the relative standard error and observing the consequent p-value. A small p-value will indicate a significant coefficient; otherwise, a large value will indicate a non-significant coefficient, which means that the associated variable is not a significant cause of the probability that $Y=1$ (of default in the example).

We can also assess the significance of the whole model. To this aim, we can compare the model with those that provide the best and worst fit. The first corresponds to the so-called "saturated" model, a model which, having as many parameters as observations, provides a complete description of the data and a perfect fit. The worst model is the null one with a single parameter which is the intercept. On the one hand, the null model leaves all the variability of the response variable unexplained. On the other hand, the saturated model has n parameters, one for each observation, and, therefore, attributes all the variability to the systematic component.

In practice, the null model is too simple, and the saturated model is not informative because it does not summarize the observations parsimoniously but simply reproduces them completely. However, the saturated model provides a benchmark for measuring the goodness of fit of an intermediate model with p parameters, in terms of its distance from the saturated model.

We now present the output of the considered example found by Python for the evaluation of logistic regression models. Table 10-3 reports, for each coefficient, the relative estimate, standard error, value of the test statistic (which is approximately normal, hence the symbol z), and p-value.

Table 10-3. *Estimates from a full logistic regression model*

<table>
<tr><th></th><th>coef</th><th>std err</th><th>z</th><th>P>|z|</th><th>[0.025</th><th>0.975]</th></tr>
<tr><td>const</td><td>-1.8327</td><td>0.043</td><td>-42.541</td><td>0.000</td><td>-1.917</td><td>-1.748</td></tr>
<tr><td>Interest_rate</td><td>0.3008</td><td>0.054</td><td>5.582</td><td>0.000</td><td>0.195</td><td>0.406</td></tr>
<tr><td>Installment</td><td>-0.1589</td><td>0.042</td><td>-3.770</td><td>0.000</td><td>-0.241</td><td>-0.076</td></tr>
<tr><td>Annual_income</td><td>0.0570</td><td>0.039</td><td>1.452</td><td>0.146</td><td>-0.020</td><td>0.134</td></tr>
<tr><td>dti</td><td>0.0681</td><td>0.035</td><td>1.973</td><td>0.048</td><td>0.000</td><td>0.136</td></tr>
<tr><td>Fico</td><td>-0.9630</td><td>0.064</td><td>-14.994</td><td>0.000</td><td>-1.089</td><td>-0.837</td></tr>
<tr><td>Days_of_delay</td><td>-0.0591</td><td>0.039</td><td>-1.519</td><td>0.129</td><td>-0.135</td><td>0.017</td></tr>
</table>

The output in Table 10-3 shows that the variables "Annual income" and "days of delay" do not significantly influence the probability of default, unlike the others. The next table compares the deviance of a logistic regression model with four variables (the significant ones in the previous table) with a complete six-variable model.

***Table 10-4.** Estimates from a reduced logistic regression model*

	coef	std err	z	P>|z|	[0.025	0.975]
const	-1.8298	0.043	-42.592	0.000	-1.914	-1.746
Interest_rate	0.3009	0.054	5.583	0.000	0.195	0.406
Installment	-0.1422	0.038	-3.707	0.000	-0.217	-0.067
dti	0.0603	0.034	1.766	0.077	-0.007	0.127
Fico	-0.9708	0.064	-15.211	0.000	-1.096	-0.846

	Model	Df	Log-Likelihood	Deviance	Chi2	p-value
0	Reduced	4.0	-2800.787094	5601.574187		
1	Full	6.0	-2799.096043	5598.192087	3.3821	0.184326

Table 10-4 shows that the p-value for comparing the two models is equal to 0.18 (the deviance difference is equal to only 3.38), and therefore we do not reject the hypothesis that says the two models are the same. They are indeed equivalent from the statistical viewpoint.

Tree Models

Let's now compare the logistic regression model with a tree model for the same data.

While linear and logistic regression models first produce a score and then, possibly, a classification, according to a discriminant rule, decision trees first produce a classification of the observations into groups and then a score for each of them, constant in each group.

Tree models can be defined as a recursive procedure, through which a set of n statistical units is progressively divided into groups, according to a divisive rule that aims to maximize the internal homogeneity of the groups obtained. At each step of the procedure, the divisive rule is specified by a partition of the values of one of the explanatory variables. Therefore, at each step, choosing a divisive rule involves choosing which explanatory variable to use and how to partition it.

The main result of the tree analysis is a final partition of the observations: to achieve the latter, it is necessary to specify a stopping criterion for the dividing process. Assume that a final partition has been reached, consisting of g groups (g < n).

Then, for any given response variable yi, a regression tree produces an estimated value that is equal to the mean of the response variable of the group to which observation i belongs. Let m be such a group; formally, we have that:

$$\hat{y}_i = \frac{\sum_{i=1}^{n_m} y_{lm}}{n_m}.$$

For a classification tree, however, as seen for logistic regression, the estimated values are given in terms of estimated probabilities of belonging to a single group. Assume that only two categories are possible (binary classification); the probability of the event can take the values 0 or 1 considering a threshold, and, therefore, the estimated probability corresponds to the observed proportion of successes in group m. Figure 10-1 represents the result of the tree models for the example considered.

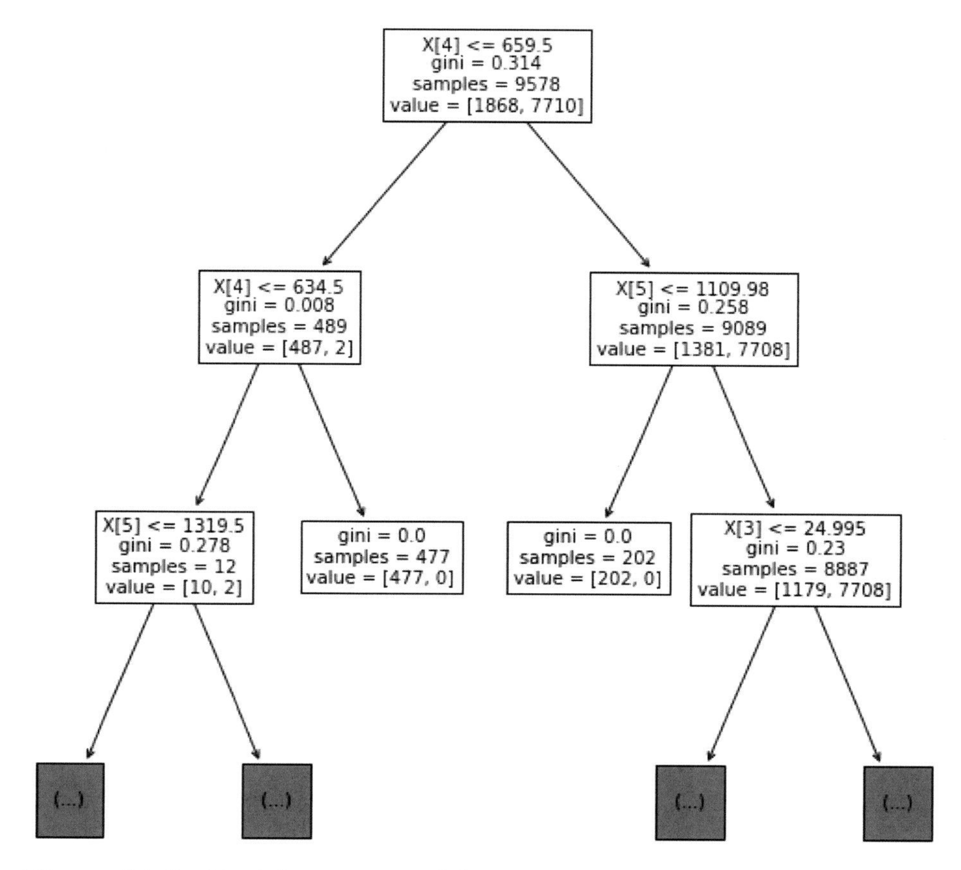

Figure 10-1. *Decision tree model*

Figure 10-1 shows that the sample is first divided based on the values of the variable X4 (debt to income ratio). On the left, the elements of the sample with X4 less than 659.5 are further divided, again on the basis of X4 and, then, in terms of X5 (FICO score). On the right, the elements of the sample with X4 greater than 659.5 are divided in terms of X5 (FICO score) and, then, in terms of X3 (log income).

Classification trees can be unstable, that is, they can be very dependent on the data considered. To this end, they are evolved into "random forests" that estimate multiple trees corresponding to different samples and average the related results. In other words, the random forest estimates are means of the values estimated from the individual trees.

$$\frac{1}{B}\sum_{1}^{B}\hat{y}_i$$

The problem with random forest models is that, being averages between multiple trees, they do not allow us to understand the causal factors that enable us to identify the path which leads to a certain estimate. To this end, we consider the so-called "variable importance plots" which indicate how much each variable reduces the variability of the original sample. Figure 10-2 shows an example for the data considered.

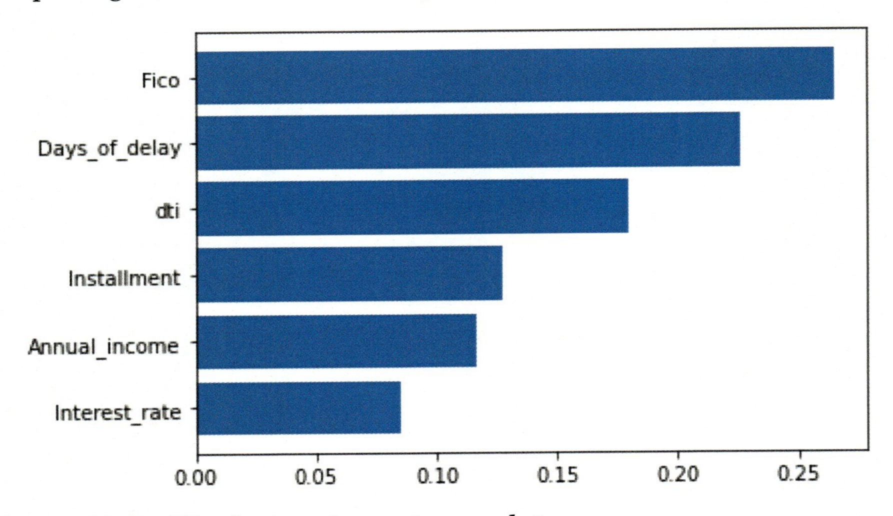

Figure 10-2. *The feature importance plot*

Figure 10-2 identifies FICO score, days of delay, and dti as the most important variables.

Neural Networks

Neural networks were originally developed to try to mimic the neurophysiology of the human brain through combining simple computational elements (neurons) into a highly interconnected system. They have become an important, and probably the best known, computational method of data mining.

A neural network is composed of a set of elementary computational units, called neurons, connected to each other by means of weighted connections. These units are organized into layers such that each neuron in a layer is exclusively connected to the neurons of the previous and subsequent layers. Each neuron (also called node, or unit) represents an autonomous computational unit and is reached (in input) by a series of signals that can determine its activation. Following activation, each neuron produces an output signal. All input signals reach the neuron simultaneously such that the neuron receives more than one input signal, but transmits only one output signal.

For each input signal, a connection weight is associated which determines the different relative importance that the input signals can have in producing the final impulse emitted by the neuron. Connections can be excitatory, inhibitory, or null (i.e., absent) depending on whether the corresponding weights are positive, negative, or null, respectively. The weights are adaptive coefficients which, in analogy with the biological model, are modified in response to the various signals traveling on the network based on an appropriate learning algorithm. Furthermore, a threshold value called bias is usually introduced. Its role is similar to that of an intercept in a regression model. Neural networks are black box models: they may be very accurate, providing accurate predictions, but their output is not interpretable.

Model Comparison

The classification models introduced in this chapter - logistic regression, tree (and forest) models, and neural networks - are of different nature: some have an underlying probabilistic model (the former); others do not (the second and third). They are therefore not comparable using statistical tests. Instead, they can be compared directly in terms of the accuracy of their predictions.

If, in the predictive phase, a model correctly classifies more individuals than another (e.g., as reliable or unreliable debtors), it will be considered better.

This way of reasoning can be formalized in the concept of a confusion matrix. The confusion matrix contains the number of elements classified correctly or incorrectly for each class. On its main diagonal appears the number of objects classified correctly for each class, while the extra-diagonal elements indicate the number of objects classified incorrectly. If it is (explicitly or implicitly) assumed that each incorrect classification has the same cost, the proportion of incorrect classifications out of the total cases constitutes the error rate and represents the quantity that must be minimized. Clearly, the assumption of equal costs can be replaced by errors weighted with the relative costs.

Table 10-5 shows an example of a confusion matrix. The table classifies the observations of the test dataset into four possible categories: (1) observations predicted as events and actually such (with an absolute frequency equal to a); (2) observations predicted as events and actually not events (with frequency equal to c); (3) observations predicted as non-events and actually events (with frequency equal to b); (4) observations predicted as non-events and actually such (with a frequency equal to d).

Table 10-5. *Theoretical confusion matrix*

Predicted/Actual	Reliable (1)	Not Reliable (0)	Total
Reliable (1)	a	b	a+b
Not Reliable (0)	c	d	c+d
Total	a+c	b+d	a+b+c+d

For a binary classification problem such as the credit one considered in this chapter, the confusion matrix is as in Table 10-6.

Table 10-6. *Observed confusion matrix*

Predicted/Actual	Reliable	Not Reliable
Reliable	45	30
Not Reliable	15	60

In Table 10-6, the main diagonal shows the correct predictions. It can be seen that the model predicted 60 borrowers as reliable (creditworthy). Of these, however, only 45 were actually reliable (i.e., they repaid the credit granted on time), while 15 proved unreliable. Similarly, 90 debtors were predicted as unreliable, but, of these, only 60 were actually unreliable. The overall accuracy rate of the model is therefore equal to 105/150 = 70%.

We note that, for reasons of efficiency and confidentiality, the above forecasting exercise is usually carried out on a so-called test sample, rather than on real data. To this end, the available data sample is divided into a training sample, which is used to learn the model (training set), and a validation sample, whose Y data are used as actual values to be compared with the predicted values.

Another important observation concerns the transition from logistic regression to the confusion matrix. Logistic regression does not provide predictions in class form (zero and one) but in score (probability) form. It is therefore necessary to "round" the probabilities to zero or one. The rounding threshold is arbitrary. It is not always correct to use 50% (majority rule) as a threshold. To solve this problem and have a less arbitrary and more agnostic criterion for comparing models, we use the ROC (receiver operating characteristic) curve.

The ROC curve is a graph that also measures the predictive accuracy of a model. It is based on the confusion matrix (Table 10-5).

Given an observed table and a value on the basis of which to carry out the rounding (cutoff), the ROC curve is calculated on the basis of the resulting joint frequencies of predicted and observed events (reliable individuals) and non-events (unreliable individuals).

More precisely, it is based on the following conditional probabilities:

- **Sensitivity (a / (a+b))**: Proportion of events predicted as such
- **Specificity (d / (c+d))**: Proportion of non-events, predicted as such
- **False positives (1-specificity) (c / (c+d))**: Proportion of non-events, predicted as events (type II error)
- **False negatives (1-sensitivity) (b / (a+b))**: Proportion of events predicted as non-events (type I error)

The ROC curve is obtained by representing, for each fixed cutoff value, a point on the Cartesian plane which has the false positive value (1-specificity) on the x-axis and the true positive value (sensitivity) on the y-axis. Each point in the curve therefore corresponds to a particular cutoff.

In terms of comparing models, the best curve is the one most markedly shifted to the left; ideally, it should coincide with the y-axis. The area under the curve synthetically represents the performance of a model. It is between zero and one; the closer to one, the better the predictive ability of a given model. Figure 10-3 provides an example of an ROC curve and calculation of the AUROC (Area Under the ROC).

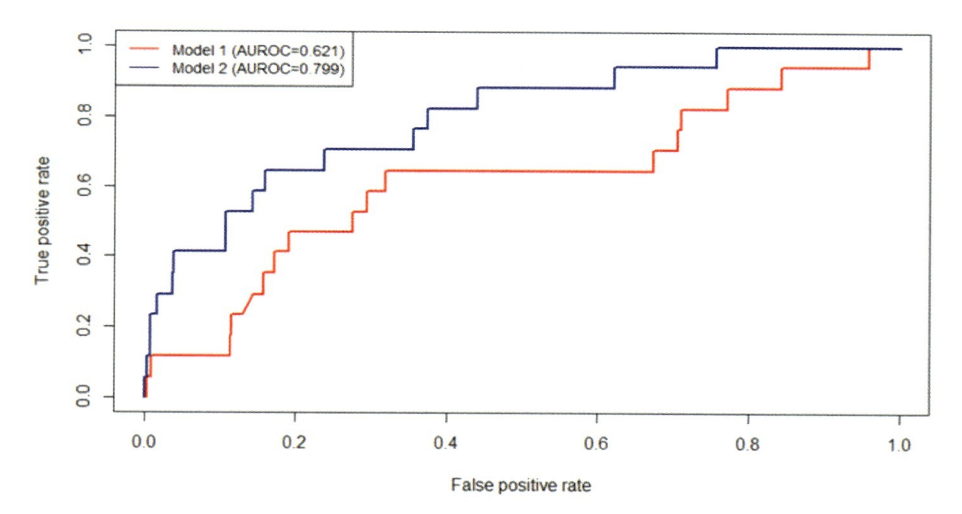

Figure 10-3. *ROC (receiver operating characteristic)*

From Figure 10-3, it shows that the best model is Model 2, with an AUROC equal to 0.799 against 0.621 of Model 1.

We can now compare the different scoring models seen in the chapter. Let's start with logistic regression.

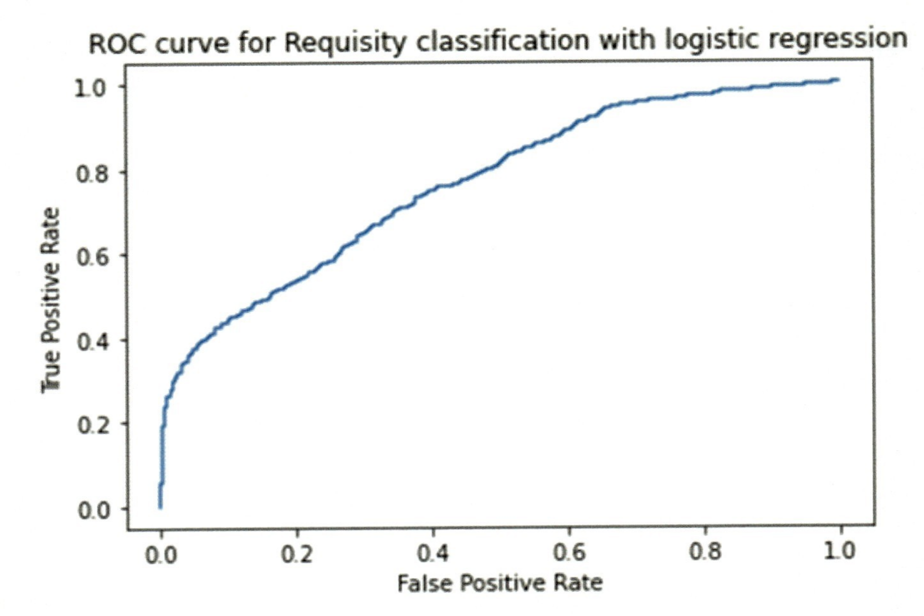

Figure 10-4. *ROC curve for the logistic regression model*

The logistic ROC represented in Figure 10-4 has an underlying area equal to AUROC = 0.7612. Regarding tree models: the tree model in Figure 10-5 has an AUROC equal to 0.70987, therefore worse than logistic regression.

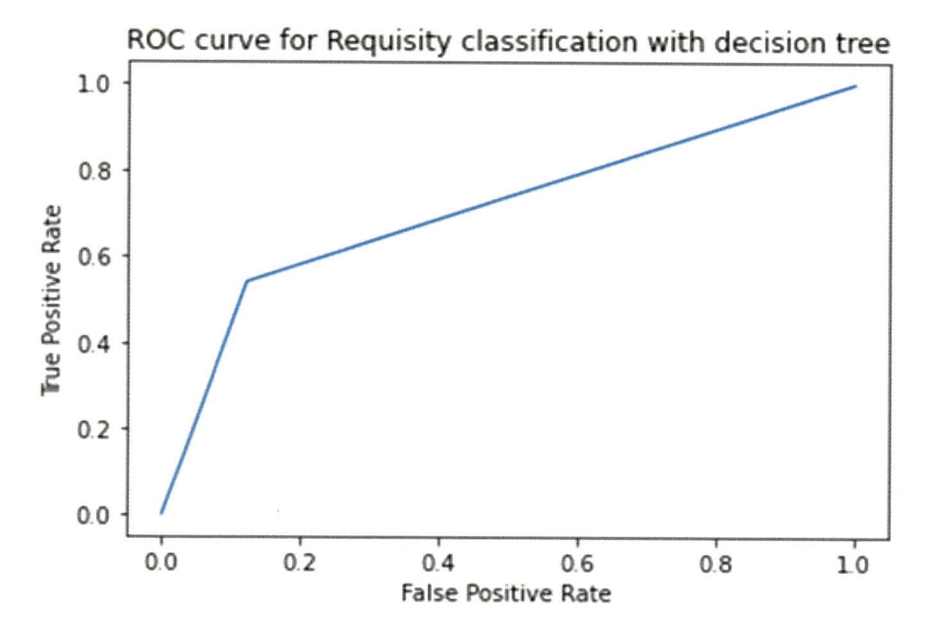

Figure 10-5. *ROC curve for the decision tree model*

Let us now consider the random forest model.

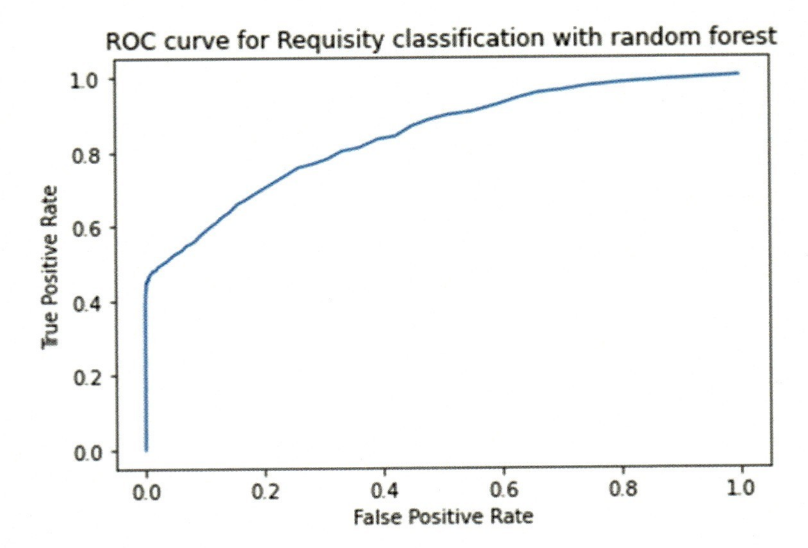

Figure 10-6. *ROC curve for the random forest model*

In this case, the AUROC is equal to AUC = 0.8390, better than the three models considered so far.

Finally, let's consider the neural network.

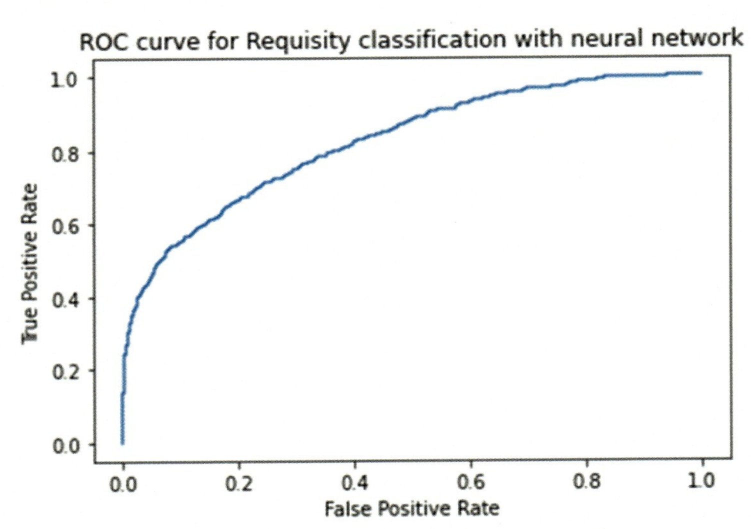

Figure 10-7. *ROC curve for the neural network model*

For the neural network, AUROC = 0.8204. The result is very good, but slightly lower than that of the random forest, which is therefore the best model.

SAFE-HAI Assessment

The wide use of machine learning models requires developing risk management models that can balance opportunities with risk of using these models. "safeaipackage" is a Python package which implements the SAFE-HAI approach proposed in this book and has a wide range of built-in capabilities to implement the metrics introduced in the previous chapters. The capabilities are divided into a set of modules, each centered around the measurement of different AI principles: accuracy, security, explainability, fairness, privacy, sustainability. This package can be installed using `pip` as follows (more information about this package is available on `https://github.com/GolnooshBabaei/safeaipackage`):

```
pip install safeaipackage
```

Here are more details on the functions described in the Python package applied to the considered data in this chapter.

Accuracy

To find the accuracy of the models, we can use the "check_accuracy" module in which we have the "Accuracy" class. Using this module, we can find the RGA metric for all the models we used in the previous section. Remember that, for a binary response, RGA=AUROC.

Logistic regression:

```
acc_logistic_reg = check_accuracy.Accuracy(xtrain, xtest,
ytrain, ytest, model_lr)
rga_logistic_reg = acc_logistic_reg.rga()
```

```
rga_logistic_reg
>>> 0.5925286280058539
```

This output says that the accuracy of the model considering the rga function is equal to about 0.59, showing a not accurate model.

Decision tree:

```
acc_tree = check_accuracy.Accuracy(xtrain, xtest, ytrain,
ytest, model_tree)
rga_tree = acc_tree.rga()
rga_tree
>>> 0.7097687795788048
```

We see that the accuracy of the decision tree model is about 0.71, better than the logistic regression model.

Random forest:

```
acc_rf = check_accuracy.Accuracy(xtrain, xtest, ytrain, ytest,
model_forest)
rga_rf = acc_rf.rga()
rga_rf
>>> 0.730511801466253
```

Doing the same for the random forest model, we find out that the accuracy is better than the previous two models, equal to about 0.73.

Neural network:

```
acc_nn = check_accuracy.Accuracy(xtrain, xtest, ytrain, ytest,
model_network)
rga_nn = acc_nn.rga()
rga_nn
>>> 0.6826493361169493
```

The accuracy of the neural network model is almost equal to the decision tree model, almost equal to 0.7.

Explainability

To evaluate the explainability of the the models, we can use the "check_explainability" module, in which there is the `rge()` function that can calculate the RGE metric for all the explanatory variables in each model.

Logistic regression:

```
exp_logistic_reg = check_explainability.Explainability(xtrain,
xtest, ytrain, ytest, model_lr)
rge_logistic_reg = exp_logistic_reg.rge()
rge_logistic_reg
>>>

RGE
Fico              0.326453
Interest_rate     0.083154
Installment       0.054892
Days_of_delay     0.036232
dti               0.024276
Annual_income     0.018660
```

Looking at the results found by the explainability module for the logistic regression model, we can see that "FICO" has the highest contribution, while "Annual_income" shows the smallest RGE value.

Decision tree:

```
exp_tree = check_explainability.Explainability(xtrain, xtest,
ytrain, ytest, model_tree)
rge_tree = exp_tree.rge()
rge_tree
>>>
RGE
Fico             0.338447
Days_of_delay    0.303410
```

```
dti              0.270506
Interest_rate    0.264183
Annual_income    0.233464
Installment      0.213933
```

We see that again here "FICO" is the most important variable.

Random forest:

```
exp_rf = check_explainability.Explainability(xtrain, xtest,
ytrain, ytest, model_rf)
rge_rf = exp_rf.rge()
rge_rf
>>>
RGE
Fico             0.199945
Days_of_delay    0.097260
dti              0.091773
Interest_rate    0.046991
Annual_income    0.025638
Installment      0.016505
```

As it can be seen, the results of the random forest are almost similar to the decision tree but different from the RGE values found for the logistic regression model. For example, while "Installment" is the third significant variable, it is the least important variable for the random forest model.

Neural network:

```
exp_nn= check_explainability.Explainability(xtrain, xtest,
ytrain, ytest, model_nn)
rge_nn = exp_nn.rge()
rge_nn
>>>
RGE
```

```
Fico            0.249193
Interest_rate   0.163634
Days_of_delay   0.136245
dti             0.124837
Installment     0.093381
Annual_income   0.081396
```

Robustness

To evaluate the robustness of the models, we can use the "check_robustness" module. Using this module, there are two approaches to measure the robustness. Using the `rgr_all()` function, we can check the robustness of the model when all the variables are perturbed at the same time. Another option is the `rgr_single()` function which checks the robustness of the model when perturbing only one variable. Here, we report the results of the `rgr_all()` function.

Logistic regression:

```
rob_lr= check_robustness.Robustness(xtrain, xtest, ytrain,
ytest, model_lr)
rgr_lr = rob_lr.rgr_all()
rgr_lr
>>> 0.4907575208408846
```

Decision tree:

```
rob_dt= check_robustness.Robustness(xtrain, xtest, ytrain,
ytest, model_dt)
rgr_dt = rob_dt.rgr_all()
rgr_dt
>>>  0.519388902090104
```

Random forest:

```
rob_rf = check_robustness.Robustness(xtrain, xtest, ytrain,
ytest, model_rf)
rgr_rf = rob_rf.rgr_all()
rgr_rf
>>> 0.47538129620810476
```

Neural network:

```
rob_nn = check_robustness.Robustness(xtrain, xtest, ytrain,
ytest, model_nn)
rgr_nn = rob_nn.rgr_all()
rgr_nn
>>> 0.46748678355614437
```

Looking at the results of the four models, we can see that the highest RGR is for the decision tree model, which states that this model is the most robust classifier in our example.

Fairness

To check the fairness of the models, we can use the "check_fairness" module available in the package to evaluate the group fairness based on the protected variable. Here, we consider the "Annual income" as the protected variable. To this end, we group the observations to "low_income" and "high_income" considering the average value of the income.

Logistic regression:

```
fair_lr= check_fairness.Fairness(xtrain, xtest, ytrain, ytest,
model_lr)
rgf_lr = fair_lr.rgf(["Annual_income"])
rgf_lr
>>> 0.926443
```

Decision tree:

```
fair_dt= check_fairness.Fairness(xtrain, xtest, ytrain, ytest,
model_dt)
rgf_dt = fair_dt.rgf(["Annual_income"])
rgf_dt
>>>  0.922399
```

Random forest:

```
fair_rf = check_fairness.Fairness(xtrain, xtest, ytrain, ytest,
model_rf)
rgf_rf = fair_rf.rgf(["Annual_income"])
rgf_rf
>>> 0.972109
```

Neural network:

```
fair_nn = check_fairness.Fairness(xtrain, xtest, ytrain, ytest,
model_nn)
rgf_nn = fair_nn.rgf(["Annual_income"])
rgf_nn
>>> 0.924922
```

Looking at the RGF values, random forest model represents the highest level of fairness toward the protected variable.

To conclude this chapter, we've reviewed a case study for credit loan applications and looked at various AI models that could be used during this analysis. We also provided a brief practical introduction on the SAFE-HAI Python package and how it relates to the mentioned AI models.

We hope you have gained valuable insights from reading this book. As AI technologies are steadily being integrated to existing tech infrastructure and are a part of people's daily workflows, it's crucial that the development of these technologies are supported by practical and implementable guidelines on responsible, ethical, and safe ways to build AI systems that pose less risks to users and society. Building on the Responsible AI framework introduced by Duke, already mentioned in Chapter 1, we've expanded the framework, as a second step to include accuracy, security, and sustainability. We also paid special attention to human rights which is a fundamental challenge facing many people from minority groups with their interaction with AI. These additions form the SAFE-HAI framework which is the core focus of this book. We extensively demonstrated and explained ways to measure each principle across each chapter in this book, with key metrics using a "rank graduation" score. This is a new and novel way to measure Responsible AI principles for any AI model or application.

AI systems cannot be properly developed, deployed, and used without appropriate governing processes in place. To this end, we provided a brief review on the various existing AI policies and ways to include AI governance in your organization. We've wrapped up the book, with an in-depth case study, looking at the process for evaluating a credit loan application using the SAFE-HAI framework.

As AI technologies and use are on the rise, let's ensure Responsible AI processes and practices are also on the rise. Otherwise, we'd have a big problem on our hands with a proliferation of technologies that increasingly pose more harm than good to members of society, consumers, and the world at large. We hope you've gained valuable insights to develop Responsible AI practices to your AI/ML life cycle and workflows.

Appendix

Chapter 1

1. Neal Johnson, Responsible AI – an overview, www.hippodigital.co.uk/responsible-ai-an-overview.
2. Jakob Uszkoreit, Transformer: A Novel Neural Network Architecture for Language Understanding, www.blog.research.google/2017/08/transformer-novel-neural-network.html.
3. Toju Duke (2023). Building Responsible AI Algorithms, Apress.
4. AIAAIC, LAION-5B image-text pairing dataset, www.aiaaic.org/aiaaic-repository/ai-algorithmic-and-automation-incidents/laion-5b-image-text-pairing-dataset.
5. Chloe Xiang, A Photographer Tried to Get His Photos Removed from an AI Dataset. He Got an Invoice Instead, www.vice.com/en/article/pkapb7/a-photographer-tried-to-get-his-photos-removed-from-an-ai-dataset-he-got-an-invoice-instead.

T. Duke and P. Giudici, *Responsible AI in Practice*,
https://doi.org/10.1007/979-8-8688-1166-1

6. Benj Edwards, Artist finds private medical record photos in popular AI training data set, www.arstechnica.com/information-technology/2022/09/artist-finds-private-medical-record-photos-in-popular-ai-training-data-set.

7. Audrey Azoulay, Towards an Ethics of Artificial Intelligence, www.un.org/en/chronicle/article/towards-ethics-artificial-intelligence.

8. UNESCO, Recommendation on the Ethics of Artificial Intelligence, www.unesco.org/en/articles/recommendation-ethics-artificial-intelligence.

9. United Nations, Final Report – Governing AI for Humanity, www.un.org/ai-advisory-body.

10. The White House, Executive Order on the Safe, Secure and Trustworthy Development and Use of Artificial Intelligence, www.whitehouse.gov/briefing-room/presidential-actions/2023/10/30/executive-order-on-the-safe-secure-and-trustworthy-development-and-use-of-artificial-intelligence.

11. EU Artificial Intelligence Act, The EU Artificial Intelligence Act, www.artificialintelligenceact.eu.

12. UK Parliament, Artificial Intelligence (Regulation) Bill, www.bills.parliament.uk/publications/53068/documents/4030.

13. Paolo Giudici and Emanuela Raffinetti (2023). SAFE Artificial Intelligence in finance, Finance Research Letters, Volume 56, 104088, ISSN 1544-6123.

14. BIS, The Basel Framework, www.bis.org/basel_framework/index.htm.

15. Max Lorenz (1905). Methods of measuring the concentration of wealth. Publications of the American Statistical Association, 9(70): 209–219.

16. Corrado Gini (1921). Measurement of Inequality of Incomes. Economic Journal, 31, 124–126.

17. Global Citizen, The Richest 1% Own Almost Half the World's Wealth & 9 Other Mind-Blowing Facts on Wealth Inequality, www.globalcitizen.org/en/content/wealth-inequality-oxfam-billionaires-elon-musk.

Chapter 2

18. Raffinetti, E. (2023). A rank graduation accuracy measure to mitigate artificial intelligence risks. Quality and Quantity, 57, 131–150.

19. Giudici, P. and Raffinetti, E. (2024). RGA: a unified measure of predictive accuracy. Advances in Data Analysis and Classification.

20. Huang, Y., Zhang, Q., Yu, P., Sun, L. (2023). TrustGPT: A Benchmark for Trustworthy and Responsible Large Language Models.

21. Diebold, F. and Mariano, R. (1995). Comparing Predictive Accuracy, Journal of Business and Economic Statistics, 13 (3), 253–263.

22. DeLong, E.R., DeLong, D.M., and Clarke-Pearson, D.L. (1988). Comparing the areas under two or more correlated receiver operating characteristic curves: a nonparametric approach. Biometrics, 44(3), 837–845.

23. Wang A., Pruksachatkun Y., Nangia N., Singh A., Michael J., Hill F., Levy O., Bowman S. (2020). SuperGLUE: A Stickier Benchmark for General-Purpose Language Understanding Systems.

24. Jeffrey Ip, An Introduction to LLM Benchmarking, `www.confident-ai.com/blog/the-current-state-of-benchmarking-llms`.

25. AWS, Amazon Machine Learning – Developer Guide, `www.docs.aws.amazon.com/pdfs/machine-learning/latest/dg/machinelearning-dg.pdf#improving-model-accuracy`.

Chapter 3

26. Croce, F., Andriushchenko, M., Sehwag, V., Debenedetti, E., Flammarion, N., Chiang, M., Mittal, P., Hein, M. (2023). Robustbench: a standardised adversarial robustness benchmark, arXiv paper.

27. James, G., Witten, D., Hastie, T., Tibshirani, R. (2023). An introduction to statistical learning, with applications in Python and R. Springer-Verlag.

28. Babaei, G., Giudici, P., and Raffinetti, E. (2024). SafeAIpackage: a python package for AI risk management. SSRN.

29. Babaei, G. and Giudici, P. (2024). GPT classifications, Machine learning with applications.

30. Bai, Y., Anderson, B.G., Kim, A., Sojudi, S. (2023). Improving the accuracy-robustness tradeoff of classifiers via adaptive smoothing. arXiv:2301.12554.

31. Gridlex, Building Robust Machine Learning Models: Tips and Tricks, `www.gridlex.com/a/building-robust-machine-learning-models-st6324`.

Chapter 4

32. Bracke, P., Datta, A., Jung, C., and Shayak, S. (2019). Machine learning explainability in finance: an application to default risk analysis. Staff Working Paper No. 816, Bank of England.

33. XAI-2024, The 2nd World Conference on eXplainable Artificial Intelligence, `www.xaiworldconference.com/2024`.

34. Shapley, L. (1953). A value for n-person games. Contributions to the Theory of Games, II, 307–317, Princeton University Press.

35. Lundberg, S. M. and Lee, S. (2017). A unified approach to interpreting model predictions. Proceedings of the 31st International Conference on Neural Information Processing Systems (NIPS), 4768–4777.

36. Giudici, P. and Raffinetti, E. (2021). Shapley Lorenz explainable artificial intelligence. Expert Systems with applications, 167, 114104.

37. Saranya, A., Subhashini, R. (2023). A systematic review of Explainable Artificial Intelligence models and applications: Recent developments and future trends, Decision Analytics Journal, 7.

38. Babei, G., Giudici, P., Raffinetti, E. (2023). Explainable fintech lending. Journal of economics and business, volume 125–126.

Chapter 5

39. Teodorescu, M. and Yao, X. (2021). Machine learning fairness is computationally difficult and algorithmically unsatisfactorily solved. IEEE High Performance Extreme Computing Conference (HPEC), 1–8.

40. Horesh, Y., Haas, N., Mishraky, E., Resheff, Y., Meir Lador, S. (2020). Paired-consistency: An example-based model-agnostic approach to fairness regularization in machine learning. Machine Learning and Knowledge Discovery in Databases: International Workshops of ECML PKDD 2019, 590–604.

41. Karimi, H., Khan, M.F.A., Liu, H., Derr, T., and Liu, H. (2022). Enhancing individual fairness through propensity score matching. IEEE 9th International Conference on Data Science and Advanced Analytics, 1–10.

42. Grabowicz, P. A., Perello, N., and Mishra, A. (2022). Marrying fairness and explainability in supervised learning. ACM Conference on Fairness, Accountability, and Transparency, 1905–1916.

43. Jiang, H. and Nachum, O. (2024). Identifying and correcting bias in machine learning. Google research paper.

44. Agarwal, S., Muckley, C., and Neelakantan, P. (2023). Countering racial discrimination in algorithmic lending: A case for model-agnostic interpretation methods. Economics Letters, 26.

45. Chen, Y., Giudici, P., Liu, K., Raffinetti, E. (2024). Measuring fairness in credit ratings, Expert Systems with Applications, Volume 258, 125184.

Chapter 6

46. Terzi, E., Liu, K. (2010). A framework for computing the privacy scores of users in online social networks. ACM Transactions on Knowledge discovery data, 5:6, 2010.

47. Li, X. (2020). Graph convolutional networks for privacy metrics in online social networks. Social networks in applied sciences, 10, 2020.

48. Babaei, G., Giudici, P., and Raffinetti, E. (2024). Safeaipackage: a Python package for AI risk management, SSRN.

49. Medium, Privacy-preserving machine learning: Techniques for protecting sensitive data, www.medium.com/@zhonghong9998/privacy-preserving-machine-learning-techniques-for-protecting-sensitive-data-d199b450e5a9#:~:text=Homomorphic%20encryption%20is%20a%20cryptographic,even%20during%20the%20computation%20process.

Chapter 7

50. United Nations, Governing AI for Humanity, www.un.org/sites/un2.un.org/files/un_ai_advisory_body_governing_ai_for_humanity_interim_report.pdf.

51. Morgan Stanley (2019). ESG approaches and principles (Report). Morgan Stanley. www.morganstanley.com/im/publication/resources/esg-approach-and-principles-enja.pdf.

52. Forstater, M., & Zhang, N. (2016). ESG definitions and concepts: Background note. UNEP Inquiry: Nairobi, Kenya.

53. Delmas, M. A., & Burbano, V. C. (2011). The drivers of greenwashing. *California management review*, 54 (1), 64–87.

54. Furlow, N. E. (2010). Greenwashing in the new millennium. *The Journal of Applied Business and Economics*, 10 (6), 22.

55. Strobel, G. (2020). Making sense of ESG: A primer on social corporate responsibility (Forbes, Ed.). `www.forbes.com/sites/forbesfinancecouncil`.

56. Agosto, A., Cerchiello, P., Giudici, P. (2023). Bayesian learning models to measure the relative impact of ESG factors on credit ratings. International journal of data science and analytics.

57. Bosone, C., Bogliardi, S. M., Giudici, P. (2022). Are ESG Female? The hidden benefits of female presence on sustainable finance. Review of Economic Analysis, 2022, 14(2), pp. 253–274.

58. Ali, S., Liu, B., & Su, J. J. (2018). Does corporate governance quality affect default risk? The role of growth opportunities and stock liquidity. *International Review of Economics & Finance*, 58, 422–448.

59. La Rosa, F., Liberatore, G., Mazzi, F., & Terzani, S. (2018). The impact of corporate social performance on the cost of debt and access to debt financing for listed European non-financial firms. *European Management Journal*, 36 (4), 519–529.

60. Forbes, We Need To Make Machine Learning Sustainable. Here's How, `www.forbes.com/sites/esade/2023/03/17/we-need-to-make-machine-learning-sustainable-heres-how`.

Chapter 8

61. Giudici, P., and Raffinetti E., (2023). SAFE Artificial Intelligence in Finance. Finance Research Letters, 56, 104088.

62. Giudici, P., Abu-Hashish, I. (2019). What determines bitcoin exchange prices? A network VAR approach. Finance Research Letters, vol. 28, 309–318.

63. Abercrombie, G., Benbouzid, D., Giudici, P., Golpayegani, D., Hernandez, J., Noro, P., Pandit, H., Paraschou, E., Pownall, C., Prajapati, J., Sayre, M., Sengupta, U., Suriyawongful, A., Thelot, R., Vei, S., and Waltersdorfer, L. (2024). A Collaborative, Human-Centred Taxonomy of AI, Algorithmic, and Automation Harms.

64. Giudici, P., Piergallini, A., Recchioni, M., and Raffinetti, E. (2024). Explainable AI for time series models. Physica A.

Chapter 9

65. Bommasani, R., Kapoor, S., Klyman, K., Longpre, S., Ramaswami, A., Zhang, D., Schaake, M., Ho, D., Narayanan, A., Liang, P. (2023). Considerations for governing open foundation models, Issue Brief HAI Policy & Society.

66. IBM, What is spear phishing? `www.ibm.com/topics/spear-phishing#`.

67. Anderljung, M., Smith, E., O'Brien, J., Soder, L., Bucknall, B., Bluemke, E., Schuett, J., Trager, R., Strahm, L., Chowdhury, R. (2023). NeurIPS 2023.

Chapter 10

68. Babaei, G., Giudici, P., Raffinetti, E. A Rank Graduation Box for SAFE AI, Expert Systems with Applications, Volume 259 (2025).

Index

A

T. Duke and P. Giudici, *Responsible AI in Practice*,
https://doi.org/10.1007/979-8-8688-1166-1

B

C

D

E

H

I

J, K

L

R

S

T

U

V

W, X, Y, Z

Printed in the United States
by Baker & Taylor Publisher Services